# More enough?

**An optimistic assessment of world energy**

**Edited by Robin Clarke**

**The Unesco Press**

The authors are responsible for the choice and
the presentation of the facts contained in this
book and for the opinions expressed therein, which
are not necessarily those of Unesco and do not
commit the Organization.

The designations employed and the presentation
of material throughout this publication do not
imply the expression of any opinion whatsoever on
the part of Unesco concerning the legal status
of any country, territory, city or area or of its
authorities, or concerning the delimitation of
its frontiers or boundaries.

Published in 1982 by the United Nations
Educational, Scientific and Cultural Organization
7 place de Fontenoy, 75700 Paris
Printed by Imprimerie des Presses Universitaires
de France, Vendôme

ISBN 92–3–101986–4

Illusrations by
Dorothy Christie and
Małgorzata A. Pióro

# Preface

More Than Enough? An Optimistic
Assessment of World Energy *is the first
published title to appear in Unesco's new
*SEXTANT *series—books designed to bring the
specialists' thought on pressing human and
societal issues to a wide spectrum of readers,
from the man in the street to the
decision-maker. Few issues are more
pressing today than that of energy—man's
needs for energy and his efforts to satisfy those
needs. This first *SEXTANT *book brings
together writers who have examined this issue
from a variety of viewpoints.*

# Publisher's note

Most of the chapters in this book are adapted or lightly revised versions of articles that have appeared elsewhere.

Chapters 4, 9, 11, 13 and 16 are reproduced unabridged from one issue of Unesco's quarterly publication, *Impact of Science on Society*, Vol. 29, No. 4, October-December, 1979, this issue appearing under the theme 'The Energy Potential'.

Chapters 5, 7, 10 and 12 are edited versions of selected chapters from the document, *The Environmental Impacts of Production and Use of Energy*, Nairobi, United Nations Environment Programme, 1980. (Report of the Executive Director.)

Chapters 2, 14 and 15 are edited versions of selected sections of a document by Alan McDonald, *Energy, A Problem of the 80s, of the 90s, of the 21st Century*, Laxenburg (Austria), International Institute for Applied Systems Analysis, 1981, which summarizes the results of a seven-year study conducted at the IIASA and reported completely in a two-volume book, *Energy in a Finite World*, Cambridge, Mass., Ballinger Publishing Co., 1981.

Unesco has dipped into its files of unpublished documents or manuscripts to round out the book's coverage (Chapters 1, 3 and 6).

# Contents

# Foreword

by Robin Clarke

Whatever else may be said about the 'energy crisis' of 1973, it was no crisis. The *Concise Oxford Dictionary* defines 'crisis' as 'time of danger or suspense'. What happened in 1973 was not a moment of danger but the beginning of a long-term trend, a sudden realization that the energy supply so much taken for granted for so many years was finite. From here on, we have a problem. It is not one that will go away.

In the 1970s many people saw this as the beginning of the end. It was a time of doom and gloom, and the computer predictions of that period nearly all emphasized the fact that civilization appeared not to be sustainable for very far into the future. But computers are only as good as their operators; and very often they are used merely to confirm the self-fulfilling prophecies of their programmers. If we now look to the future, ten years after the crisis began, we begin to see the shape of a different world. During the past decade many have sat back and taken stock. And they have found that there is no energy shortage as such. True, the one fuel—oil—on which we have relied too heavily in the past will be in short supply in the future. But other fossil fuels such as coal will last a good while yet. And the as yet untapped energy available in a myriad of other sources is copious. It is not energy we are short of: it is technology, time and money.

Others might add that we may also be short of wisdom, for the crucial problem now is to evolve a strategy which will lead us as smoothly as possible towards a future which is sustainable over the long term. How are we to do it?

This book looks at the future of our energy problem from two points of view, which I have called the hard options and the soft options. It must be emphasized that these are not alternatives. If the world of tomorrow is going to be satisfactory in any sense, we cannot opt simply for one or the other; we must have both. Each of these two options encapsulates a different way of looking at the world and its problems. And these two different approaches are not restricted to questions of energy, they are found almost anywhere that what is perceived as a 'global problem' exists. Let us look first at the hard options.

The hard-option approach is characterized by words such as 'global', 'overall', 'macro-scale' and 'average'. The facts it considers run as follows. The world population today is something over 4,500 million people. They consume energy at a rate of slightly more than two kilowatts per head. In round numbers this means our global energy use is at the rate of 10,000 million kilowatts. This is the same as

9

$10^{12}$ watts, which in scientific shorthand is called 10 terawatts or 10 TW. This is our yardstick.

Short of a global catastrophe on an unprecedented scale, nothing will stop the world population reaching 6,000 million by the year 2000. More probably it will by then reach 6,500 million or more. And while per capita energy consumption in the industrialized world may not rise appreciably in the next two decades, it must do so in the developing world. Today three-quarters of mankind consume less than two kilowatts per head, some 400 million of them less than 100 watts. The average energy consumption in the developing countries is only 450 watts.

If the per capita energy consumption rises to three kilowatts in the next twenty years, global energy demand will be at least 18 TW, probably 20 TW, twice what it is today. All global experts expect that both population and energy consumption will level off during the following century, with a 'final' demand of somewhere between 30 and 100 TW. If that amount of energy is to be found in the future—and it is between three to ten times the amount we use today—it turns out that we shall have to rely to a much greater extent on three possible options: nuclear fission, nuclear fusion and solar power. These, then, are the hard options. Again, no one pretends that we must simply choose any one of the three. They will be used in varying combinations and in combination with continuing but diminishing supplies of fossil fuels.

Each has its own problem. Nuclear fission, as it is currently practised, is very wasteful of fuel. Indeed, the uranium needed to power present-day reactors would run out early in the next century unless we also use the more productive nuclear breeder reactor. That, however, involves difficult, and some would say dangerous, technologies that are by no means mastered as yet. Breeder reactors also produce plutonium, which is not only one of the most toxic materials known, but is the basis of nuclear weapon production. While many people might argue that this is a technically feasible route into the future, nearly everyone would agree it is also the most hazardous.

Nuclear fusion, relying primarily on an almost infinite supply of deuterium fuel in seawater, could solve our energy problems for all time. The problem here is that no one has yet been able to make controlled fusion work in a sustainable way. The research effort is now massive. Even so, no one expects to see working fusion power-plants until well into the next century.

Solar energy is also an appealing solution. As everyone now knows, the problem here is not the availability of the energy source, which is potentially enormous, but the fact that it is so diffuse. To harness it usefully is not even technically very difficult, but it is expensive, whether it is done by massive arrays of photo-electric cells, or by thousands of large reflectors tracking the sun and reflecting its heat onto a central boiler used to raise steam and turn a turbine. Solar energy also has one advantage which neither fission nor fusion can claim. Solar power could be used to run a 1,000 megawatt power-station; it could also be used to heat an individual house, power an irrigation pump or light up a village. Solar energy, in other words, is flexible in scale; fission and fusion are not.

In connection with solar energy, we cannot leave the zone of hard options without a brief mention of photochemical energy conversion—in particular, the photolysis of water, which involves large-scale production of hydrogen. As with the other hard options, however, the process promises to be expensive, yet photoconversion should be operative by the mid-1980s.

These possibilities bring us to the soft options, characterized by words such as flexible, partial, rural, local and appropriate. The hard options are concerned with global energy demand, and finding the resources to match them on the macroscale. The soft options are concerned with smaller resources, and with smaller pockets of demand. It is no good providing sufficient global energy to meet a theoretical demand if the million villages of the developing countries are still without power.

With the exception of solar energy, the soft options alone cannot possibly meet global demand. They include hydro-, geothermal and wind power, energy from the oceans and from biomass, and also solar energy used on a smaller scale. None of these by itself is ever likely to produce more than say 5 TW (half of today's global energy use), and most a great deal less. But the fact that they do not represent a global solution does not imply they should not be used. On the contrary there are perhaps twenty sites in the world where efficient and useful tidal power-plants could be built. No one would suggest they should not be built because they cannot provide solutions elsewhere in the world. In the world of tomorrow we must make use of energy wherever it is economically feasible.

The soft options have other important characteristics. Hydropower, for instance, can be used to provide light for a hamlet with a generator of no more than two kilowatts. It can also be used for a 100-megawatt power-station. Its ability to operate on the small scale is particularly valuable, for it now seems unlikely that the West's solution of rural electrification via a national transmission network will ever be universally adopted in the Third World. The distances involved, and the capital costs, are simply too high. By 1971 only 12 per cent of the population of the developing countries were connected to a national grid. The solutions of the future are likely to come mainly from local sources of renewable energy, which include biomass, wind, solar energy, hydropower and in some cases geothermal resources. Geothermal energy is an important resource, one which in scale might merit inclusion under the hard options. Estimated total recoverable resources are many times greater than all the fossil fuels put together.

Energy is like food. There is absolutely no doubt that starvation on the earth is quite unnecessary. Global agricultural production, even from currently exploited agricultural land, is perfectly capable of feeding 4,500 million people entirely adequately. The problems do not lie in the global statistics but in the fine details of everyday life, which range in scale from international policy to the flash flooding of a small stream in a remote hamlet in Nepal. Politics and distribution are the main keys. So, too, with energy. To be sure, we must know that resources exist to satisfy demand. They do, or can be made to. Getting the energy resources to the right people at the right price is just as important a problem but a great deal more difficult. This is why we need both hard and soft options. But we need other things too.

The greatest growth in energy consumption during the 1960s and 1970s was in electricity, a form of energy in which about three-quarters of the energy used to build and run a power station is lost. Private cars running on cheap petrol have tended to replace less energy-intensive forms of transport. Most of the energy used in the production of packaged foods is expended *after* the food leaves the farm. And so on. The fact that this book does not concern itself with matters of energy conservation does not mean that the subject is ignored; rather it is regarded as a necessary prerequisite for the proper use of the energy sources discussed herein.

Similarly, another factor frequently overlooked is the need to deliver such energy resources in the right form. For when we use the word energy we should apply a caveat. The energy we are concerned with is that which is capable of doing useful work; and useful, in practice, frequently means concentrated. That is to say, we need power to run our civilizations, and power (the rate of doing work) means having our energy in conveniently concentrated packages, such as fuel or chemical feedstocks.

Neither has there been space in this book to examine several other aspects of matching energy resources to demand which do not depend on the resources themselves. They depend on methods of using the resources. Today we are profligate in the way we both make and use our energy. We had become accustomed, until quite recently, to take cheap energy for granted. We can no longer afford to do so.

And so, finally, we need strategies. We need to plan, to organize, to look ahead. There is the matter of a long-term commitment to development of new sources of energy. But meanwhile, as an interim measure, it is essential to limit demand by appropriate conservation measures. Three sections of this book are based on a report by the International Institute for Applied Systems Analysis called *Energy in a Finite World.* It took seven years to compile this report and involved more than 140 scientists from twenty countries. It is without doubt the most authoritative statement in print on the future of our energy problem. IIASA's comments on how they see energy demand being met in the future, and on how we may reach a sustainable energy situation in the next century, are therefore of great relevance.

It is not intended to detract from the importance of those statements by giving the 'Last Word' (page 171) to someone who is not an energy specialist. Very often the affairs of men are conducted within an environment so restricted that the picture they paint is incomplete. In considering the hard options, IIASA does not give detailed consideration to the fusion option because it is technically still unsolved. Similarly, Tom Mikkelsen does not consider the fission/breeder alternative because he considers it too dangerous. His recipe for a satisfactory energy future is simple enough. It involves two things: sharing out our remaining fossil fuels fairly; and diverting our military research and development resources to the investigation of fusion and solar power.

Mikkelsen's reasoning is based on a long-term consideration of human evolution and the needs of *Homo sapiens.* The human species in its time has wrestled with mighty problems: pestilence, famine and war have taken the highest tolls. But we have come through. We shall also undoubtedly learn to manage the continuing energy crisis of the coming decades. In doing so the limits we will confront will not be those of the resources themselves so much as the limits of our own ingenuity and wisdom.

# A historical perspective on the energy problem

1

James McDivitt

*The Director of Unesco's Division of Technological Education and Research, James McDivitt, reviews the history of man's long struggle with resource restrictions and concludes that the present energy situation is a transition process rather than an emergency response to a one-time crisis. This process challenges man to use his reason in evolving a long-term policy on resource use, including energy.*

## An encounter with resource restrictions

For a very long time man has known that the earth's energy resources are not unlimited and that allowing population growth to rise rapidly along with ever-growing demands on these resources places him on a serious collision course. Yet throughout this long time, man has continued to prosper and to advance thanks to a fortuitous combination of techno-logical ingenuity and improved understanding of the true potential of the earth. From Malthus, who first stated in 1798 that growth of population tends to outrun growth of pro-duction and that poverty was the inescapable fate for most of mankind, to the 1972 Club of Rome study, *The Limits of Growth*, which provided a statistical base for the contention that drastic changes must be made in the social and economic systems if they were to survive into the middle of the next century, to the 1980 *Global 2000 Report* to the President of the

United States, which provides a sober assess-ment of the world during the next two decades, during which it must come to terms with a rapidly changing resource balance, man has been warned that at some stage and at some time the growth cycles had to be moderated. Given that there is real validity to these argu-ments, what do they mean in the case of a specific situation at a specific time?

At the present time energy is the most prominent feature in this picture, in part be-cause energy is so all-pervasive in human activities and plays such an important role in maintaining the 'quality of life' many of us have come to enjoy, in part because of the rapid plunge from a long period of ample supply of low-cost easily recovered petroleum products to a period of uncertain supply at very much higher costs. However, energy is by no means the only feature that must be considered: the non-renewable minerals, particularly some of the critical metals such as chromium and molybdenum, agricultural land, water and

15

ultimately even living space put restrictions, if not limits, on the growth of society as the privileged of the world now know it.

But this should be nothing new, for indeed society is an ever-changing thing, and these restrictions have always been with us. In Malthus's day famine and pestilence and war, each of which is to some extent a reflection of lack of available resources, kept the population at a modest level, while fifty years ago these forces were still a strong factor in limiting burgeoning populations in countries like India and China and even today we cannot say that they have been eliminated. It is often overlooked by people in the advanced world of today that significant changes have taken place in their life-style even in the course of the last twenty-five years. By the same token we should recognize that similar changes will occur over the next twenty-five years. Of course, we must also keep in mind that we who live in the relatively advanced zones and enclaves, and have fully participated in these changes, constitute only a small fraction of the total world population.

## A changing energy mix

Using energy as the example, two hundred years ago the major sources of energy were probably human and draught-animal power supplemented by firewood and animal dung for cooking and heating. To a great extent this is still true for most people in the world today. By the end of the nineteenth century coal had replaced firewood in most industrial and many transportation uses, but the world relied on a mixture of sources, still with a high degree of reliance on firewood and charcoal and on

human and animal power. It is only during the period since 1920 that petroleum and, to an increasing extent, natural gas have taken a position in which, because of what must be considered abnormally low costs and prices, they have come to replace, for industrial countries at least, the earlier mix or balance in which different sources of energy were used for different purposes. During this period, for example, draught animals were replaced by tractors, watermills and windmills by electric motors, coal- and wood-burning stoves and furnaces by oil and gas: in each case petroleum or natural gas taking over from a traditional source. But from the beginning of the petroleum era engineers and economists have recognized that there were limits and that the ultimate limit for the non-renewable resources such as the mineral fuels was the stock existing in the earth which cannot be expanded and which therefore must serve all people for all time. In this case all that we can say is that these limits, which may approach the absolute twenty, fifty or a hundred years from now, are more critical today than they were twenty or fifty years ago, and that, following the patterns of the past, man, assisted by technology, must once more change the structure of his energy mix.

It is useful and helpful to look upon this as what it really is, part of a continuing transition which has been going on for centuries and which must continue throughout the history of mankind, not only in energy but in all resource-related areas. On this basis one might foresee for energy an accelerating shift during the remaining years of the century from oil and natural gas towards other known sources of energy for which technology exists, hydrocarbons such as coal, oil shale and tar-sands, nuclear fission, wind and geothermal energy. As we move into the next century we can expect that some of these sources will begin to taper off as the supply of electric power serving the growing urban complexes is progressively taken over by sustainable energy sources including breeder reactors, large-scale solar and other renewable forms (ocean, geothermal, etc.) and possibly by nuclear fusion. It is far more accurate to look upon the present energy situation as part of a transition process rather than as an emergency response to a one-time crisis, although there are elements of emergency in the way that the world is reacting.

In fact today's pressures are induced by price rather than by supply, since reserves of petroleum and natural gas are still almost at record levels, new reserves are being discovered and proven each year, and it can be certain that these energy sources will continue to play a major role into the indefinite future. However, it is equally certain that they cannot continue to meet almost 70 per cent of the world's commercial energy needs as they have done during the last decade, and that we are approaching the stage where they must taper off not only in percentage but also in absolute quantities as the largest and most accessible reserves are used up. It is highly unlikely that we will ever run out of oil and natural gas (or of coal or uranium for that matter) but the market mechanism will assure that dwindling and ever more costly supplies will be used in those high-value products such as petrochemicals and fine lubricants, for which there is no ready substitute. Today's use of petroleum as a fuel to generate electricity denies a future generation access to much more valuable products that are difficult to obtain from other sources.

## Reason and wisdom needed . . .

Petroleum serves well to illustrate what is happening and what must happen as population grows and the demand on limited resources increases, and, indeed, there are very valuable lessons to be learned from the present energy situation and mankind's reaction to it. For at long last we appear to be approaching the point in time when we must revise our approach to natural resources—a revolution that has been creeping up on us for some generations with the growing awareness of limits as reflected in the growing interest in conservation, ecology and the environment, but which will take its most definite form in the recognition of the need for a world policy for the use of resources. In simplest terms this implies nothing more than wise use giving due account to the needs of all people who live on the earth today and who will live on it in the future, but this will not be easy, particularly since it must be rather quick. These concepts underlie much of the discussion at the Law of the Sea Conference, which illustrates well the conflicts and problems involved.

In man's eyes the earth has always been an object for plundering, and resources, whether they be forests, land, water or mineral fuels, have always been exploited in a reckless and profligate way. Attitudes and approaches have got to change, but it is difficult to expect the habits and traditions of a hundred generations to be wiped out in the remarkably short time that concepts of conservation and of so-called 'limits to growth' have been with us. Fifty years ago few people even in the most developed circles entertained such thoughts, and even today relatively few really understand the implications, even the medium-term implications, of the continuation of these attitudes of the past. This being true in a developed society, consider the confusion in the mind of a peasant in a traditional subsistence society whose whole philosophy of life is based on the fact that the earth will provide (for example firewood for fuel) when it becomes apparent to him that today the earth is having difficulty providing, and that tomorrow the supplies will not be enough. In such a situation we can be reasonably sure that the rich and developed countries will find a way to purchase survival, but this must not be at the expense of the poor and underdeveloped.

## . . . beginning with energy use

This is the situation in which the world finds itself in the case of energy, serious but not critical, provided we grasp the true problem and deal with it as a problem for all mankind and for all time. In fact, the current energy situation provides a wonderful opportunity for a rational, harmonized approach to a fundamental challenge, and if mankind can face it on this level, it can point the way to dealing with other resource-related issues that will arise in the foreseeable future. On this level, the fact that the challenge comes first in energy is a happy circumstance, since energy is a fundamental resource that can be used directly or indirectly as a substitute for many other resources, and dealing adequately with energy will, to some extent, ease the pressure on other natural resources.

It is this type of challenge which Unesco has been preparing for throughout its relatively brief history across the wide front of resource-related fields. Again, concerning energy, in 1948 Unesco organized in India a meeting on solar energy within the framework of its arid-zones programme which was one of the actions leading to the establishment of the International Solar Energy Society, with which close working links are still maintained. Over a long period when alternative energy was either a scientific curiosity or a hobby, Unesco and other United Nations bodies continued with programme activities, limited it is true, but still enough to keep the subjects warm and, more important, to introduce many hundreds of scientists from developing countries to these subjects. In 1973, on the eve of the awakening of economic interest in alternative sources of energy, Unesco hosted an international congress, 'The Sun in the Service of Mankind', which defined the state-of-the-art for solar energy at that key moment in time.

Since then, continuing efforts have been made within the United Nations system to provide a base for understanding the medium- and long-term implications of the present energy situation, culminating in the United Nations Conference on New and Renewable Sources of Energy, in Nairobi in August 1981. The question these activities lead up to is not how to cope with the present so-called energy crisis—if that is the question we have lost the opportunity mentioned above. The question must be how to develop and share resources, using energy as a model, so that all mankind, both today and in the indefinite future, may benefit from this, the heritage of all mankind. As with many such questions there may be no answer to satisfy everyone, but simply posing the question forces a different approach, and simply seeking an answer to this question will lead to a different and probably more acceptable interim solution.

The present volume highlights the fact that for energy there is nothing to fear from the resource side and that indeed the problems are in how we develop and use what nature still provides, for although man is a plunderer the world can withstand unbelievable abuse, and even after centuries of plundering it remains filled with treasures. But the limits which yesterday may have seemed at infinity are today visible as as a speck on a distant horizon, and can no longer be ignored.

To quote one well-known authority: 'We are *not* running out of energy. Ninety-nine per cent of all energy that will *ever* be available for human use is in the sunlight that strikes the earth. Humankind's current commercial energy budget equals only 0.01 per cent of this solar influx; a hundred-fold increase would equal only one per cent.'[1] This illustrates wonderfully the illusive nature of the resource problem for the future, for to catch a sunbeam is not so easy. In the case of energy, there are many alternatives, and the sun, the ultimate source, has no practical limits. It seems likely that we have time; it seems likely that we have the scientific and technical ability and skills; have we learned enough to combine these elements into a long-term solution?

1. Denis Hayes, 'Overcoming Solar Barriers: Progress in the United States', *Non-technical Obstacles in the Use of Solar Energy*, p. 2, Brussels, CEC, 1980.

# Exploring the limits

## 2

Alan McDonald

*This chapter inquires into the ultimate technical potential of various energy sources and concludes that the world's energy resources are tremendous, although taking advantage of this abundance can be neither quick nor cheap. It thus provides a clearer picture of the options ultimately available—both their good and bad sides.*

The aim of this chapter is to be imaginative, to be exploratory, to stretch our thinking. The method is to ask, for each of the different possible energy sources, what its ultimate technical potential would be if only resource constraints and limitations on technological build-up rates were considered. Problems of environmental impacts, safety questions, or mismatches between supply and demand patterns are initially assumed essentially solvable, and the constraints of competitive economics are left for Chapter 14.

The conclusion is that the world's energy resources are tremendous, although taking advantage of this abundance can be neither quick nor cheap. Exploring the implications of expanding any one energy source to the unprecedented scale necessary to supply the needs of a rapidly growing population defines vividly the associated safety and environmental ques-

tions. The purpose here is neither to determine an ideal level of use for each energy source nor to define acceptable levels of environmental impacts. It is rather to give a clearer picture of the options ultimately available—both their good and bad sides.

The presentation borrows the categories most often used in discussions of energy supply: fossil fuels, including coal, oil, and gas; nuclear power, including fission and fusion; centralized, high-technology solar power; decentralized, but not necessarily low-technology, solar power in conjunction with other renewables.

Considerations of fossil fuels usually begin with estimates of reserves and resources, and the IIASA study was no exception. Where it differed from past studies was in its concentration on unconventional resources—on deep off-shore oil, on oil available only with tertiary recovery methods, on gas in tight formations or geopressure zones, on off-shore coal deposits, on tar sands and oil shales—in short, on fossil resources much more expensive in terms of money, environmental impacts, and possible

Edited version of Section 3: 'Energy Supply: Exploring the Limits' (IIASA, 1981).

social effects than the world is traditionally used to.

Table 1 summarizes the resultant estimates of global fossil resources. The numbers in the first column, which represent the conventional fossil resources, add up to slightly more than 1,000 terawatt-years (TWyr), which corresponds well with conventional wisdom concerning global fossil resources. But in the last column, where the unconventional, expensive resources are also included, it turns out that the total is almost 3,000 TWyr, three times higher.

TABLE 1. Estimates of global fossil-fuel resources measured in terawatt-years (TWyr). The price categories are specified in barrels of oil equivalent (boe) for oil and gas and in tons of coal equivalent (tce) for coal.

| Resource | Category 1 | Category 2 | Category 3 | Total |
|---|---|---|---|---|
| Coal[1] | 560 | 1 019 | — | 1 579 |
| Oil[2] | 264 | 200 | 373 | 837 |
| Gas[2] | 267 | 141 | 130 | 538 |
| TOTAL | 1 091 | 1 360 | 503 | 2 954 |

1. For coal: Category 1, $25/tce or less; Category 2, $25–$50/tce.
2. For oil and gas: Category 1, $12/boe or less; Category 2, $12–$20/boe; Category 3, $20–$25/boe.

It is by now no surprise that coal proves to be by far the most abundant of the fossil resources. But its dominance raises two problems. The first concerns how coal is to be used to satisfy the most pressing component of the demand for fossil fuels—the liquid fuel component—and the second concerns the distribution of coal resources around the world.

In looking at the first problem, it became apparent that the coal liquefaction schemes currently being developed all rely on autothermal processes; that is, of the three basic ingredients involved in producing liquid hydrocarbons from coal—carbon, hydrogen, and heat—all three come from the coal. The alternative is an allothermal process, where the hydrogen and the heat come not from the coal but from some other source. Clearly the most important immediate effect of such an approach would be a decrease in the amount of coal needed to produce a given amount of liquid fuel. Only a quarter to one-third of the coal required by autothermal processes is needed for the allothermal schemes. But almost as important, the carbon dioxide released to the atmosphere is reduced to a quarter to one-third of the level associated with autothermal methods.

In the short term, and at the national level, these differences between autothermal and allothermal coal liquefaction are not crucial. But the world is likely to be relying on coal, particularly for the production of liquid fuels, to an increasing extent for at least the next half century. In this light, extending by a factor of three to four the portion of the world's coal resources devoted to producing liquid fuels becomes a more urgent priority.

The second point to be made about coal concerns its geographical distribution. As Table 2 shows, three countries will dominate the world coal market: China, the United States, and the USSR. The principal implications of this are clear, if coal is to replace oil as the world's principal fossil fuel: (a) the

TABLE 2. The distribution of global coal resources in thousand millions of tons of coal equivalent ($10^9$ tce)

| Greater than $10^{12}$ tce ($1{,}000 \times 10^9$ tce) | | Between $10^{11}$ and $10^{12}$ tce (100 and $1{,}000 \times 10^9$ tce) | | Between $10^{10}$ and $10^{11}$ tce (10 and $100 \times 10^9$ tce) | | Between $10^9$ and $10^{10}$ tce (1 and $10 \times 10^9$ tce) | |
|---|---|---|---|---|---|---|---|
| USSR | 4 860 | Australia | 262 | India | 57 | German Dem. Rep. | 9.4 |
| United States | 2 570 | Germany, Fed. | | South Africa | 57 | Japan | 8.5 |
| China | 1 438 | Rep. of | 247 | Czechoslovakia | 17.5 | Colombia | 8.3 |
| | | United Kingdom | 163 | Yugoslavia | 10.9 | Zimbabwe | 7.1 |
| | | Poland | 126 | Brazil | 10 | Mexico | 5.5 |
| | | Canada | 115 | | | Swaziland | 5.0 |
| | | Botswana | 100 | | | Chile | 4.6 |
| | | | | | | Indonesia | 3.7 |
| | | | | | | Hungary | 3.5 |
| | | | | | | Turkey | 3.3 |
| | | | | | | Netherlands | 2.9 |
| | | | | | | France | 2.3 |
| | | | | | | Spain | 2.3 |
| | | | | | | Dem. Peoples' Rep. of Korea | 2.0 |
| | | | | | | Romania | 1.8 |
| | | | | | | Bangladesh | 1.6 |
| | | | | | | Venezuela | 1.6 |
| | | | | | | Peru | 1.0 |

technical infrastructure required to move vast quantities of coal or coal products from the resource-rich to the resource-poor countries must be developed; and (b) the associated institutional infrastructure must be developed, for, although the current patterns of the world's balance of payments problems may shift, the problems will by no means vanish simply as a result of a global shift to coal.

For the case of nuclear power the summary also begins with resource estimates. But here there is an additional element, which arises because of the variety of nuclear technologies—which range from existing light-water reactors (LWRs) through fission fast-breeder reactors (FBRs) to fusion technologies—and the fact that the amount of energy that can be extracted from the earth's nuclear resources depends critically on whether the introduction of these technologies is co-ordinated so that they complement each other as productively as possible.

For fission reactors the resource in question is natural uranium. The estimate we arrived at for the amount ultimately available globally at prices under \$130/kg (1978 prices) was 24.5 million tons. How much energy can be produced from this amount depends on how the uranium is used.

If it is used solely to fuel LWRs and if

spent fuel is not recycled, the conclusion is that the resource could be exhausted by 2030. This estimate is based on a reference case, which assumes that additional LWRs are introduced at the highest rate still consistent with, *inter alia*, an independent assessment of the projected capabilities of the worldwide nuclear industry. This reference case led us to a nuclear power production level of 17 TWyr/yr (thermal) in 2030 and, as mentioned above, the exhaustion of the world's high-grade natural uranium resources by the same date.

The immediate question is, 'How may the lifetime of nuclear fission power be extended?' There are three possible approaches.

The first involves mining the earth's vast deposits of low-grade uranium ore; deposits that were not included in the 24.5-million-ton estimate made above. The disadvantage is that the low-grade ores—ranging from uranium concentrations of 500 parts per million (ppm) down to 30 ppm—would be much more expensive, both financially and environmentally, than the higher-grade ores. For example, Table 3 compares the land requirements, manpower requirements, and the amount of material that must be handled in order to support LWRs fuelled by 70-ppm uranium ore, with those same requirements for LWRs fuelled by high-grade ore (2,000 ppm of uranium). From the requirements for coal-powered electricity shown in Table 3 one can see that the mining requirements for the case of low-grade ore exceed those for coal.

The second approach stretches the lifetime of the high-grade uranium resources by assuming both improved efficiencies in LWRs and recycling of the nuclear fuel. But in extending our reference case along these lines, the 24.5 million tons of high-grade ores could not be made to last much more than ten to twenty years beyond 2030, even on the basis of optimistic assumptions. Afterwards, the only option is again the low-grade, expensive resources.

The third possibility is to introduce breeder reactors, the family of fission reactors capable of using the more than 99 per cent of natural uranium that cannot be used directly in LWRs. Considerations of breeder reactors usually envision a system based on LWRs of current design and an increasing proportion of breeder reactors that gradually replace the LWRs, eventually doing so altogether. The problem with this approach is that the world is already behind schedule; breeder reactors have not been and are not being developed and introduced at the necessary speed. But if the introduction of breeders is pursued in conjunction

TABLE 3. The requirements for operating a one-gigawatt (electrical) power plant

| | Land 30-year total (km²) | Mining personnel (man-yr/yr) | Material handling involved 30-year total (10⁶ tons) |
|---|---|---|---|
| LWR (2,000 ppm ore) | 3 | 50 | 45 |
| Coal | 10–20 | 500 | 321 |
| LWR (70 ppm ore) | 33 | 300 | 360 |

with enhanced LWR efficiencies, it turns out that the full potential of the breeders can ultimately be exploited. The approach that is necessary in order to reach the required improvements in LWR efficiencies assumes the gradual introduction of the uranium isotope known as uranium-233 as a fuel for LWRs. The source of this uranium-233 is presumed to be thorium-232 converted in the breeder reactors; the result is a system capable of extracting a total of 300,000 TWyr of energy from the 24.5 million tons of high-grade uranium resources.[1]

The two other obvious bases for a sustainable energy system are nuclear fusion and solar power. The commercial introduction of nuclear fusion at a global level, is, we feel, more than fifty years away; rather than speculate that far into the future here, we will simply state the energy potential of fusion and leave it at that. Deuterium–tritium reactors could tap a resource equal to approximately 300,000 TWyr,

the same as that made available by fission reactors. Introducing deuterium–deuterium reactors would enhance this estimate by a factor of 1,000, leading to a total fusion potential of 300 million TWyr.

Solar power is a more immediate possibility than fusion power, and therefore deserves more elaboration. We shall distinguish between 'hard' uses of solar energy and 'soft' uses; the label 'hard solar' refers to applications involving large centralized technologies, while 'soft solar' refers to decentralized uses on a smaller scale.

The potential of hard solar is tremendous. The average energy input to the earth from the sun is 178,000 TWyr/yr of thermal energy; even after accounting for the filtering effect of the atmosphere, the usable sunlight shining in locations suitable for hard solar technologies is sufficient to provide energy equal to hundreds of terawatt-years each year. Considering the possibility of solar plants located in space outside the earth's atmosphere increases the calculated solar potential even more. Thus, as in the case of nuclear power, solar energy can be imagined as the basis for a sustainable energy system, with the energy supply of the future independent of resource considerations for ever.

But in identifying this potential, and especially in concluding that the necessary usable land area suitable for hard solar technologies exists, two qualifications must be mentioned.

First, as in the case of fossil fuels, the world's solar resource is unevenly distributed among countries. In particular, much of the area most suitable for solar power-plant sites lies in northern Africa and the Middle East, areas already rich in oil and gas. A crucial dimension of exploiting the solar potential is

1. The 300,000 TWyr associated here with nuclear fission power is larger by a factor of 100 than the total resources of both conventional and unconventional fossil fuels (Table 1). More particularly, it is large enough to justify contemplating a sustainable global energy system based on nuclear power. But in doing so, it is crucial to remember that these 300,000 TWyr only become available if the world's uranium resources are used, not to fuel burner reactors, but to build up a system of both burner and breeder reactors, a system through which the energy supply of the future could become effectively independent of any resource considerations. Such a system we label 'sustainable', and the use of existing resources to create such a system we label 'investive'. The alternatives to investive uses of resources are the current 'consumptive' uses that characterize both existing LWRs and, necessarily, the fossil fuels.

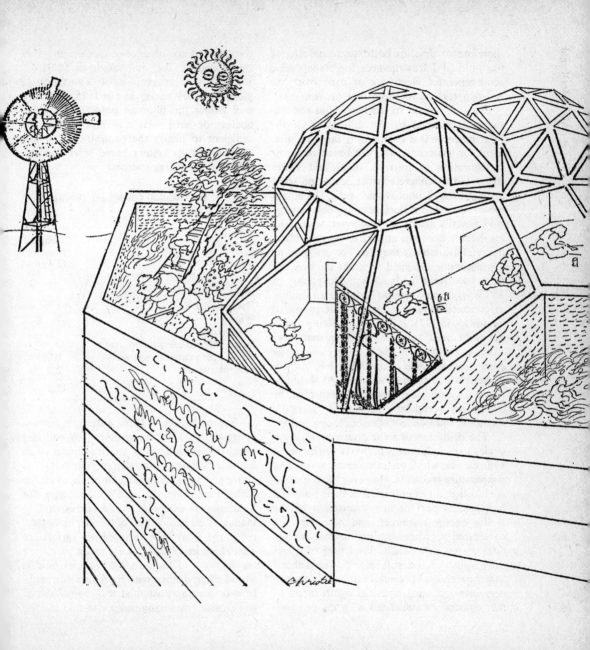

therefore to develop both the technical and institutional infrastructures for transporting solar-generated electricity or fuels from the sun-rich regions to those that are sun-poor.

Second, related to the large land requirement necessitated by the diffuseness of the solar resource is a comparably large requirement for materials; whether based on some configuration of mirrors, pipes and valves supported by concrete structures or on some arrangement of photovoltaic cells, the equipment required to collect incoming solar energy is necessarily extensive. Moreover, while land availability does not appear to be a problem, material availability may be. For orientation, a programme designed to build up over the next 100 years a hard solar capacity of 35 TWyr/yr could require each year an amount of concrete roughly equal to that produced worldwide in 1975. It is an intimidating result, but what must be remembered is that *using material resources to build up a global solar energy system would be another example of the investive use of existing resources*. As would be the case with nuclear power, the return on the investment would be a future energy supply essentially independent of resource constraints.

The definition of solar power is often extended to include energy derived from biomass, hydropower, wind, ocean currents, waves and temperature gradients. However these sources are labelled, an examination of their potential is a critical part of any assessment of the earth's energy resources, and here they are considered together with geothermal energy, tidal energy and decentralized uses of direct solar insolation, i.e. soft solar power. Table 4 lists the technical potential estimated for each (the term technical potential again indicates that constraints associated with the environ-

ment and competitive economics are not taken into account). The total shown in Table 4 is 17.2 TWyr/yr, which is more than twice the global primary energy use in 1975. Still, it is well below the ultimate potential of either nuclear or hard solar power, and is hardly sufficient to justify the possibility of a sustainable energy system based solely on this collection of energy sources.

TABLE 4. The technical potential of renewables and soft solar power

| Source | Technical potential (TWyr/yr) |
|---|---|
| Biomass | 6 |
| Hydroelectricity | 3 |
| Wind | 3 |
| Geothermal | 2 |
| Ocean thermal energy conversion | 1 |
| Tides, ocean currents and waves | 0.045 |
| Soft solar power | 2.2 |
| TOTAL | 17.2 |

But the figures in Table 4 are by no means insignificant. Most importantly, to consider using these resources at the maximum levels indicated would be to contemplate undertaking active ecological management on an awesome scale. Exploiting the 6 TWyr/yr listed for biomass, for example, would correspond to managing 30 million km$^2$ of forests, more than twice the land area devoted to agriculture worldwide in 1975. It would mean managing the habitats of thousands of species, and it would mean dealing with more familiar problems on an unprecedented scale: problems of soil erosion, managing water systems, and the decreasing pest resistance of cultured plants. In short, it would mean operating a worldwide herbarium.

The general conclusion to be drawn from the exploration of supply limits summarized here is that nuclear fission, nuclear fusion, hard solar power, or some combination of the three can provide the basis for a sustainable global energy system. The fossil fuels, soft solar technologies, hydroelectricity, biomass, and all the other energy forms considered here can play only a supplementary, though by no means an insignificant, role.

But this conclusion is based on looking far into the future; and to identify where the world could end up in perhaps another 100 years is very different from determining the direction in which it is headed now. What might we expect during the next fifty years? Only after this question has been answered can we try to predict what a transition from the world's current energy system to a sustainable energy system might actually look like.

# The hard options

# Power from fission

3

Z. Zaric

*Nuclear fission now provides about 6 per cent of the world's electricity and may double by 1985. For the present, the principal nuclear power plant is the thermal reactor, which, however, suffers the disadvantage of a very low utilization of uranium: only about 1 per cent of the potential energy in natural uranium. The nuclear breeder reactor promises distinctly higher utilization factors, perhaps as high as 60 to 70 per cent, but not before solutions are found to such serious problems as disposal of highly radioactive wastes from the fuel-reprocessing plants, possibilities of large and destructive transients, dangers in handling sodium metal, etc. There is also a political problem to be met, namely the opposition of the general public to fast-breeder plants.*

Nuclear energy is already extensively used in large commercial nuclear power-plants for the production of electrical power. More than 200 industrial-scale nuclear reactors are already in operation, with a total installed capacity of more than 120 gigawatts (GW) electrical. About one-third of these reactors are in operation in the United States, supplying about 12 per cent of the electricity produced in that country. Of the other countries, Japan, USSR, France, the United Kingdom, the Federal Republic of Germany, Canada and Sweden each have more than ten reactors in operation. A few developing countries, such as India, the Republic of Korea and Pakistan also have nuclear power-plants in operation. About 6 per cent of all the electrical energy produced in the world is now of nuclear origin.

Installed nuclear capacity is expected to double by 1985, as a result of the exponential growth-rate of installed capacity during the 1970s. In some European countries, such as France, it is expected that at least 50 per cent of electrical production in the late 1980s will be of nuclear origin. The main reason for this rapid growth is that, for instance in France in 1978, nuclear power was considered 40 per cent cheaper than electricity from oil and 20 per cent cheaper than electricity from coal. The capital investment needed for a nuclear power-plant is, however, considerably higher than in a fossil-fuel plant due to greater complexity and to the fact that it currently

Unpublished manuscript from Unesco files submitted by Professor Z. Zaric, International Centre for Heat and Mass Transfer, Belgrade, Yugoslavia takes about ten years to build a nuclear power-station. On the other hand, in 1978, nuclear fuel was three times cheaper per kilowatt-hour of electricity than oil.

## Thermal reactors

The overwhelming majority of nuclear power-plants in operation and being built are thermal reactors. Of these, three out of four are of the light-water type (LWR), either pressurized-water (PWR) or boiling-water (BWR) type. The technology of light-water reactors was derived from the development of the nuclear-submarine power reactor in the United States. LWRs employ 2 to 4 per cent enriched uranium. Of the reactors fuelled with natural uranium only heavy-water moderated and cooled reactors (HWRs), being developed in Canada, are currently employed in commercial power-plants, but on a much smaller scale. These reactors are also favoured in some developing countries, such as India, mainly because they do not require fuel enrichment. Natural-uranium (graphite), gas-cooled reactors were used in the United Kingdom and France for a certain period, but could not compete with enriched-fuel reactors due to high investment costs. Enriched-fuel, graphite-moderated reactors are employed in the United Kingdom, with gas cooling, and in the USSR with water or steam cooling.

Nuclear power-plants are currently less efficient than modern fossil-fuel power-plants. The overall efficiency of a current PWR nuclear power-plant is around 33 per cent, that of a heavy-water (HWR) nuclear power-plant is even lower at about 29 per cent. Graphite-moderated reactors have substantially higher overall efficiencies. Current power-plants being built in the United Kingdom employ so-called 'advanced gas-cooled reactors' (AGRs) which use enriched uranium and $CO_2$ cooling, and reach efficiencies comparable to those of fossil-fuel power-plants. In the USSR graphite-moderated reactors are cooled by steam, which is superheated in the reactor so that efficiencies approaching 40 per cent are achieved. Very high efficiencies are expected from the so-called 'high-temperature gas-cooled reactors' (HTGRs) which are still at a development stage. All components inside such reactors, including fuel elements, are made of ceramic material capable of withstanding high temperatures. These reactors are cooled by helium, an inert gas, and outlet temperatures from 900 to more than 1,000 °C are in principle achievable. These temperatures would lead to a high overall efficiency.

## Low utilization of uranium

The greatest disadvantage of currently employed thermal reactors is that their uranium utilization is very low. Fuel has to be taken out of an HWR-type reactor after it has delivered only about 7.5 megawatt-days (MWdays) of energy per kilogram. Theoretically, a kilogram of fissile material contains the energy equivalent of 1,000 MWdays. The corresponding fuel 'burn-up' of a PWR-type reactor is higher due to the enrichment, and amounts to about 33 MWdays/kg. Spent fuel from the reactor still contains 0.2 per cent $U^{235}$ in the case of an HWR, and 0.9 per cent in the case of an LWR. It also contains similar amounts of fissile plutonium. In the LWR case, uranium consumption is in fact not less than in the case of an HWR. Additional $U^{235}$ for the enrichment has to be provided from natural uranium in special enrichment plants. To obtain 1 kg of 3.2 per cent enriched fuel it is necessary to use as much as about 6.5 kg of natural uranium. Therefore, natural uranium consumption in an LWR is about equal to that in an HWR. As a consequence, in both reactor systems only about 1 per cent of the potential energy in natural uranium is used. This low utilization of uranium resources is typical of the nuclear power-plants presently in use.

Estimates of global nuclear power-plant capacity by the end of the century can only be speculative. They range from 2 to 3 TW electrical. With only a 1 per cent utilization of uranium resources, this means that annual natural uranium requirements will amount to 400,000 to 500,000 tons in the year 2000. Cumulative natural uranium requirements would then amount to between 3 to 4 million tons by the end of the century. From Table 1 it can be seen that this would mean that almost all estimated global natural uranium resources, at least those having acceptable extraction costs, would be exploited by that time. A much more efficient utilization of natural uranium is therefore required if nuclear energy is to play a significant role in the future.

## Nuclear breeder reactors

By definition a breeder reactor 'breeds' more fissile material from the fertile material $U^{238}$ and thorium than it consumes. Therefore, the uranium utilization in a breeder reactor is much more efficient than in the currently used thermal reactors. In fact utilization factors in the range of 60 to 70 per cent are considered feasible. This not only means a 60- to 70-fold increase of the energy content of uranium reserves but also leads to a negligible role for fuel costs in the overall costs of energy production. Thus, materials with a much poorer uranium content, including even sea-water,

TABLE 1. Estimated global uranium resources

| Region | Reasonably assured $10^3$ tons | Additional $10^3$ tons | Total resources $10^3$ tons | Annual production | |
|---|---|---|---|---|---|
| | | | | 1977 tons | 1990 tons |
| Africa | 430 | 140 | 6 600 | 9 300 | 20 000 |
| North America | 820 | 1 700 | 3 400 | 20 800 | 58 000 |
| South America | 60 | 90 | 3 600 | 200 | 600 |
| Asia | 160 | 190 | 2 600 | 10 | — |
| Europe | 590 | 290 | 6 000 | 3 500 | 36 000 |
| Oceania | 250 | 40 | 1 000 | 650 | 15 000 |
| Antarctica | — | — | 4 000 | — | — |
| World | 2 310 | 2 450 | 27 200 | 34 460 | 129 600 |

become potentially interesting, to the extent that nuclear energy resources become practically inexhaustible.

Nuclear breeder reactors are of a much more sophisticated construction than current thermal reactors. Uranium breeders have to operate with fast neutrons and, therefore, without moderators. As a consequence energy liberated per unit volume of the reactor core is much higher than in thermal reactors. This leads to the use of liquid metals (sodium and potassium) as reactor coolants. The fuel has to be highly enriched by plutonium so that the reactor contains a large inventory of fissile material. Nevertheless, fast-breeder-reactor technology is already in a very advanced stage of development, especially in France and the USSR. The first plants on an industrial scale are expected to be in operation by the early 1980s.

## Reprocessing spent fuel

There are, however, a number of serious problems to be solved before the large-scale utilization of fast-breeder reactors is possible. Fast breeders, utilizing highly enriched fuel, require a large inventory of fissile plutonium. This plutonium has to be provided by stocks accumulated from the operation of thermal reactors, not only for the initial charge but also during the whole initial doubling time of the reactor. Fissile plutonium has to be extracted from spent fuel elements in so-called 'reprocessing plants'. Spent fuel from a reactor is highly radioactive. Thus it has to be stored for several months in water at the nuclear power-plant until its radioactivity and rate of heat release are reduced by a factor of about 10,000. Even then the reprocessing of

the spent fuel has to be done by remote manipulation behind heavy shielding with a strict control of the leakage of radioactivity. Finally, permanent disposal of highly radioactive wastes from the reprocessing plants poses serious problems. Let us now look at some of these problems in more detail.

## Large transients

The physics of a fast-breeder reactor is different from the physics of thermal reactors, in the sense that rapid, transient increases of power are more likely in such reactors. In addition, the radioactive inventory of a fast-breeder reactor is much greater than the inventory of a thermal reactor of the same power. Thus, the potential hazard to the environment following a major release of radioactivity would be much greater. Therefore, the design philosophy of the safety system has to be changed. If a large transient occurs and is followed by a most unlikely simultaneous failure of two independent shut-down systems, core melting and disassembly inside the pressure vessel would occur. The reactor vessel, primary piping and the containment are designed to withstand pressure-loads corresponding to such a maximum credible accident. Therefore, the reactor would be destroyed but the radioactivity not allowed to escape. Presently operating prototypes of fast-breeder reactors are designed on these principles. In addition there is a large experimental programme in all countries developing fast-breeder technology aimed at testing all components and safety systems for the large commercial power-plants under construction.

## Dangers in handling sodium metal

High power-density and the physical characteristics of fast-breeder-reactor cores imply liquid metals as most convenient coolants. Other coolants, such as a gas, are also possible but it seems that the costs would be prohibitive. At present all fast-breeder reactors in operation and being built employ liquid sodium-potassium (NaK) as a reactor coolant. Sodium is, of course, a very reactive material. When it comes into contact with air, spontaneous combustion occurs, which is difficult to control; it also reacts explosively with water. Clearly, liquid sodium is dangerous to handle. However, this has long been known, and during the past twenty years or so a safe sodium-handling technology has been constantly developed. Normally, sodium is allowed to come into contact with inert gases only. A direct contact of the radioactive primary-loop sodium with the water in the turbine cycle is avoided by an intermediate sodium loop in which the sodium is not radioactive. Safety features against sodium fire are also provided.

Plutonium is the principal fuel of fast-breeder reactors. The initial inventory of a liquid, metal-cooled, fast-breeder reactor for a power-plant of 1 GW electrical amounts to several tons of plutonium. This plutonium is obtained from uranium in thermal or fast-breeder reactors. It has to be separated from other fission products in special fuel-reprocessing plants. Highly radioactive waste from the reprocessing plant consisting mainly of fission products has to be safely stored.

## Radioactivity hazards

Plutonium is $\alpha$-radioactive with a half-life of about 20,000 years. Very small amounts of plutonium (less than one microgram), if introduced into the body, could be fatal. In addition plutonium is an extremely dangerous explosive, and is the raw material for the manufacture of nuclear weapons. A global energy economy based on fast-breeder reactors would involve the large-scale circulation of tons of plutonium. Because of the complex fuel cycle, not only the reactor but also fuel reprocessing and fuel-element fabrication plants as well as the storage of waste present potential radioactivity hazards. And there is a need for the transportation of plutonium, highly radioactive spent-fuel elements and reprocessing waste from one site to another. As a consequence of theft or sabotage plutonium might be misused by terrorist groups.

Evidently the risks involved in a large-scale use of fast-breeder nuclear power-plants are considerably larger than in the case of commercial power plants operating at present. Therefore, at least at the moment, the general public appears to be opposed to fast-breeder plants, and some countries are still reluctant to introduce them on a large scale. The problems certainly exist but they could be solved given time and money. An extensive research and development programme, principally concerned with fast-breeder safety, is now under way.

# Controlled fusion: a status report

4

International Fusion Research Council

*Nuclear fusion, still in the research stage, is the process in which light nuclei collide and fuse to form heavier ones and release energy. Fusion takes place in high-temperature matter, as in the sun. Controlled nuclear fusion offers the prospect of a major new source of energy from the earth's light elements, an energy that is essentially limitless. If it is successfully developed, this source could be used for the large-scale generation of electricity.*

## Progress in physics research

Two basic conditions have to be satisfied in order to realize fusion reactors. First, the fuel has to be heated to a temperature, $T$, in the range from 50 to 500 million degrees. Secondly, the so-called Lawson criterion must be met, i.e. the product $n\tau$, of plasma number density $n$ and energy confinement time $\tau$ must exceed about $10^{14}$ cm$^{-3}$ sec. The most accessible reaction is that between the hydrogen isotopes, tritium and deuterium, which can ignite at 100 million °C, with

$n\tau = 10^{14}$ cm$^{-3}$ sec; other reactions, such as deuterium-deuterium (D-D), D-helium 3 or D-lithium 6, require higher values both of temperature and $n\tau$. The required values of $n\tau$ can be lowered by about an order of magnitude, but at the penalty of high circulating power in the power station envisaged.

The required conditions can be fulfilled, in principle, in two ways, by magnetic and by inertial confinement. In both methods, substantial progress has been made during the past ten years. Common to both is an improved understanding of physical phenom-

Originally published as 'Controlled Thermonuclear Fusion: Status Report', *Impact of Science on Society*, Vol. 29, No. 4, 1979, pp. 327–37. Contributors to the original paper from which this report is adapted include the following council members: C. Braams (Netherlands), M. Brennan (Australia), B. Brunelli (Italy), G. von Giercke (Federal Republic of Germany), K. Husimi (Japan), E. Kintner (United States), B. Lehnert (Sweden), D. Palumbo (Commission of the European Communities), R. Pease (United Kingdom), M. Trocheris (France) and B. Kadomstev (USSR). They were assisted by the following members of the International Atomic Energy Agency (IAEA): J. Phillips, A. Belozerov (scientific secretary) and H. Seligman. The report was submitted in final form to Sigvard Eklund, Director-General of the IAEA.

ena in high-temperature plasmas, arising from improved methods of experimental, theoretical and computational exploration and a wide range of experimental devices.

### Magnetic confinement

The most impressive progress towards the plasma conditions needed in a reactor has been achieved with the tokamak system. A dozen or so devices of increasing size and power have contributed to the understanding of tokamak behaviour. Substantial progress has also been made with other systems of magnetic confinement, although on a more limited scale and with less overall, world effort.

### Tokamak system

In this system the plasma is heated and confined by an electric current passed through the plasma in a toroidal chamber (shaped like the inflated inner tube of an automotive pneumatic wheel), with a strong external field to stabilize the current's flow. Following encouraging results in the late 1960s, there was considerable development of the tokamak programme and remarkable progress made since 1970.

Quantitatively, the ion temperature has been raised from 700 electron volts (eV) to about 2,000 eV by introducing secondary heating techniques. The confinement times have been improved, while higher densities and cleaner plasmas have been achieved. It was observed that the confinement time increased linearly with density. As a result, the value of the thermal insulation, measured by the $n\tau$ product, has been improved fortyfold to $10^{13}$ cm$^{-3}$ sec, so that there is now an overlap of what is achieved and what is needed for the substantial production of energy.

By the use of large computers, it is possible to simulate fairly accurately the evolution of tokamak discharges. These results make it possible to predict the performance of tokamaks within the range of present parameters, and to extrapolate beyond these to reactor conditions.

In the Princeton Large Tokamak, temperatures close to those required in fusion reactors were achieved by using neutral beam injection. With an injected power of about 2 million watts, the ion temperature was raised to 60 million °C and the electron temperature was increased without degradation of the confinement time. In the Soviet Union, the ratio of plasma pressure/magnetic field, $\beta$, has reached a peak value of about 4 per cent (5–10 per cent is needed to enable a more

# Miniglossary

*Coulomb scattering.* Scattering of charged particles by an attractive or repulsive magnetic field.

*Helical heliotron.* Device for obtaining high-temperature plasmas.

*Plasma.* A highly ionized substance (sometimes called the fourth state of matter), such as the sun, in which thermonuclear reactions can take place; the substance consists of electrons and atomic nuclei.

*Stellarator.* A generator in which ionized gas is confined in an endless tube by means of a very strong, externally applied magnetic field.

*Superconductivity.* When certain metals and alloys are cooled to close to absolute zero, their electrical resistance falls very low.

*Torr.* Unit of pressure used to measure high vacuum, equivalent to the classical millimetre of mercury.

*Torsatron.* Another device for obtaining high-temperature plasmas.

efficient use of magnetic fields necessary for an economical thermonuclear reactor). The latest value for $n\tau$, roughly $3 \times 10^{13}$ cm$^{-3}$, is three times larger than that reported in 1977 and is now within a factor of about three needed to reach the Lawson criterion (somewhat in excess of $10^{14}$ cm$^{-3}$ sec).

## Other magnetic systems

In the stellarator and related systems, containment of the plasma can be accomplished without requiring a net current in the closed plasma ring. To ensure equilibrium of the plasma, a magnetic field of complex geometry is produced by external coils. In the stellarator, the torsatron and the helical heliotron, this field is a twisted field possessing the property of a 'rotational transform'. As compared with the tokamak devices, these systems offer the favourable potential of running continuously.

There is a wide variety of stellarator and related geometries still to be studied.

## The diffuse pinch

Investigations have been carried out in a number of laboratories on the confinement of plasma in toroidal discharges stabilized by longitudinal fields which are only of the same magnitude as the fields resulting from the discharge current. The fluid stability of these systems has been demonstrated in detail and their self-stabilizing properties explained theoretically. In agreement with theory, reverse field pinches and screw pinches have shown stable confinement of plasma at the high values of $\beta$ ($>10$ per cent) needed for economic toroidal systems.

Up to now, these experiments were made in installations of small size where plasma/wall interactions restricted both the achievable elec-tron temperatures and the duration of the stable configuration.

## Open systems

Open systems have the advantage of topological simplicity, but the disadvantage of rapid loss of plasma and energy through the ends. They have practical advantages for a reactor: high values of beta, simple refuelling and exhaust of reaction products, smaller unit size, and relative simplicity of construction and maintenance. Therefore large efforts are justified to find ways to reduce end-losses and improve the energy balance. According to recent results, the most promising line is the magnetic mirror concept, already a serious candidate for a reactor.

Progress towards stable confinement of hot ion plasmas in magnetic mirror fields has been substantial. In a completely stable mirror-confined plasma, however, the loss of particles and energy along the magnetic field lines (caused by classical Coulomb scattering) may be too high for a fusion reactor. This problem may be overcome by the introduction of new features, such as (a) recirculation of the lost energy, or (b) reduction of end-losses by a more complicated geometry of the magnetic field, such as tandem mirrors and linked mirrors. These techniques are now being investigated. And, during the past several years, there has been increasing interest in long, straight magnetic systems with plasma heated by electron and laser beams.

## Inertial confinement

Fusion research based on inertial confinement is directed towards demonstrating the scientific feasibility of very rapidly heating and compressing small pellets of suitable fuel until conditions exist whereby thermonuclear fusion can occur and useful amounts of power can be produced. Such pellets may be heated to the required temperatures by means of lasers or electron or ion beams, or by magnetic compression.

### Lasers

Research on inertial confinement has produced implosions of pellets filled with deuterium and tritium. The observed production of thermonuclear neutrons has verified theoretical predictions in a satisfactory manner for low-power levels. Pellet compressions now yield about $10^9$ thermonuclear neutrons—several orders of magnitude more than those first observed only three years ago.

Rapid advances are being made in the development of high-power lasers and the associated optical engineering needed to bring this system to fruition. An encouraging result is that compression has been obtained using carbon dioxide lasers whose energy efficiency can approach that necessary ($>5$ per cent) to obtain a net yield of energy.

### Electron and ion beams

There has been recent progress in further development studies of the technology of relativistic electron beams to accomplish microexplosions. It was experimentally proved that a number of ways exist for magnetic focusing of electron beams on the surface of targets. Processes of energy absorption and transfer, furthermore, and a thousandfold compression of fusion fuel and its heating to kilovolt temperatures were verified experimentally. These measurements are confirmed by measurements of output and energy of fusion neutrons.

Intense ion beams have potential advantages over electron beams for achieving ignition of fusion targets. Since the stopping length is much less for ions than for electrons of comparable energy, lighter targets can be used. Many ideas for the use of ions are therefore being studied.

### Compression via imploding liners

Progress has also been made in the field of magnetic compression of plasma targets by means of imploding liners, though the combined experiment on plasma compression and heating by accelerated shells has not been carried out.

## Technology of fusion reactors

Since 1970 studies of envisaged fusion reactors and their engineering problems have been undertaken in most countries having programmes in fusion research, so that about 10 per cent of current efforts are devoted to this topic. By comparison with the magnitude of the technical developments needed, these endeavours are preliminary; they have served to define the main elements of the problems, to identify 'outline solutions' and to assess the potential feasibility of practical fusion reactors. They have laid the foundations for the planning of development programmes in engineering and other technology, provided a basis for costing, and identified the areas of research in plasma physics needing attention.

The work has been devoted to reactors based on the deuterium-tritium (D-T) reaction, because the physical conditions required for these reactions will be reached first, and because, for a given pressure and volume, this reaction yields the highest power output. The reaction defines some of the principal features of an electricity-generating reactor. The tritium has to be bred from lithium in a blanket surrounding the thermonuclear plasma; the heat has to be transferred by a coolant from the blanket to a generating plant driven by conventional thermal engines; and the blanket has to be supplemented by a region of biological shield in order to reduce the neutron and gamma radiation to safe values.

## Studies of magnetic systems

Most of the recent conceptual designs for magnetic fusion reactors have been based on the tokamak system. About a dozen tokamak designs are extant, covering both experimental and commercial systems. They have provided fundamental insights regarding tokamaks as fusion reactors; they have identified and defined key technological problems and assessed the potential for solving them. After a generation of studies, a convergence of solutions and a process of optimization are beginning to emerge.

Some of the studies have been done in considerable depth and detail, achieving a substantial degree of self-consistency. They give considerable confidence that a tokamak-based electrical generating system can be built. At the same time they involve suppositions which, though plausible, impinge strongly on practicality and cost factors—vital if thermonuclear power is to become a reality.

Studies of reactors based on other systems have involved greater extrapolation from present experimental achievement or theoretical understanding, and have not been so detailed. As a result, the choice of the most practical system for a fusion reactor remains open to question. It is an important task for the immediate future to pursue urgently these studies of alternative systems.

The class of 'fusion energy multipliers', whose main representative was always the open-ended mirror concept, has recently been extended to toroidal systems such as the 'two-component tokamak' and the 'fission-fusion hybrid' system, on which reactor schemes have been proposed. With the mirror concept, they have in common: an energy amplification factor in the range of 1–10, less severe Lawson conditions (a value for $n\tau$ of about $10^{13}$), possible operation on direct current, smaller size, and no need for special fuel supply arrangements.

The pioneering work on conceptual studies of a stellarator reactor was done in the early decades of fusion research. More recent studies have shown that helical external magnetic windings can be both expensive and difficult to assemble and maintain. It is therefore important that new theoretical and experimental work has identified simpler forms of external windings that are more applicable to reactor construction and have been tested experimentally. Now that the plasma physics problems of stellarators are yielding to experiment, a resurgence of reactor interest may be expected. An important need is to demonstrate stellarators operating at a value of beta in the range of 10 per cent or more, which will be the task of plasma physics for the coming years.

## Technology for magnetic systems

The problem of the D-T reactor can be classified under two main headings, those common to all magnetic confinement systems and those special to the system proposed. Let us first touch on the common problems.

### Reactor blankets

The thermonuclear plasma is surrounded by a blanket region which captures the neutrons in lithium to breed tritium, provides the primary heat-removal region of the nuclear energy and shields the magnetic coil system from nuclear radiation. A number of blanket geometries have now been considered, and breeding ratios greater than unity can readily be obtained, with a blanket thickness of the order of one metre. For the coolant, liquid lithium is attractive from the point of view of tritium breeding and heat characteristics, but has less desirable features in safety and resistance to corrosion and in reaction to the presence of strong magnetic fields.

### Tritium processing

Preliminary studies show that the quantities of tritium present in a fusion reactor can be safely processed and stored. Detailed design studies and experiments have not yet been carried out.

### Vacuum wall

In the concepts of reactor design, the vacuum container (or first wall exposed to the thermonuclear plasma) will be subjected to 30–50 million watts/yr/m$^2$ of fusion radiation energy in the form of neutrons, energetic particles and photons. Early data imply a radiation limit of 5–10 MW/yr/m$^2$, and a better material operating at the expected temperatures in a reactor must be developed.

### Divertors

Magnetic divertors are being designed to restrict contamination from the wall and to exhaust the large quantities of unburnt plasma and reaction products. A promising experimental start has been made on evaluating the performance of design concepts; although much work remains, divertors so far have worked as planned. Divertors involve difficult, special coil systems, and incorporating them into power reactors presents serious design problems.

### Radioactivity

Since 1970, detailed estimates of the neutron-induced activity of the structural material have been made. The stainless steel or nickel alloys and molybdenum seem to offer immediate prospects of combining practicality with the advantage of not very long-lived activity. Structural materials such as vanadium, titanium and aluminium alloys have been suggested. These, together with a number of alloying or non-metallic elements (silicon and carbon), offer possibilities deserving greater attention; every reduction of radioactivity in the surrounding structure offers further improvement with respect to a key environmental advantage of fusion, and eases problems of maintenance and repair. A systematic investigation of new material, specifically to minimize radioactivity and radiation damage in fusion reactors could be very fruitful.

### Magnetic fields

The design and construction of the superconducting magnetic field coils needed in many designs, although large and complex, would appear to require a straightforward engineering development programme. There is now a strong collaborative effort by several countries in the construction of large coils.

## Maintenance

A modular design of tokamak reactors has been explored to simplify repair and the replacement of parts. Techniques for the remote handling of radioactive components have to be further developed. These considerations have a major effect on a reactor's configuration and mechanical design. The approximate weight of the modules in some designs is 100 tonnes.

Next, we examine briefly some special problems.

## Injection of cold fuel

Experiments on the injection of fuel during the burn-cycle required in some reactor designs are urgently needed. A number of questions on this important problem remain open, and more definitive experiments are now under way.

## Heating

In addition to heating produced by passing an electric current through plasma, there are two main heating methods being developed: injection of neutral atoms and application of radio-frequency power. Striking progress has been made in the neutral injection heating. The further development of injected power seems to be a matter of straightforward engineering. But attention will have to be given to increase further the efficiency at high atom energies by exploiting injection concepts such as ripple diffusion, negative ions at higher energies and energy recovery from ion beams.

The achievement of these intense beams of neutral atoms has encouraged the concept of energy-multiplying systems such as the two-component reactor plasma mentioned earlier.

## Energy storage

Some fusion power reactors could require reversible energy storage in the range of several gigajoules (approximately $10^9$ BTU), delivered in a period of a fraction of a second to several seconds. Initial progress has resulted from the use of two different technologies, but substantial development work is needed if these techniques are to be applied to gigajoule storage systems.

# Some ideas for inertial confinement reactors

A conceptual laser fusion reactor would consist of a reaction cavity in which the thermonuclear energy is released from D-T reactions within a pellet, with thermonuclear burn initiated by a laser pulse. A blanket of lithium surrounds the source of fusion neutrons.

A feature of laser fusion reactors differing significantly from magnetically confined reactor concepts is the fact that the energy pulses represent substantial amounts of energy released on a very short time-scale. The minimal energy release (determined by both physical and economic considerations) is probably about 100 megajoules (roughly equivalent to 20 kg of high explosive). Although the hydrodynamic blast created by the pellet's micro-explosion can be controlled with relative ease—because the energy is carried by a small mass of high-energy particles—large stresses can result from high rates of neutron energy deposition in the blankets and structural materials. A major design problem in containing this energy is posed by the need for a low-pressure cavity in which the pellet can be heated and compressed by a laser pulse without

**45**

prohibitive laser energy loss along the beams' path; at the same time, a finite layer of blanket material surrounding the pellet is maintained.

Several conceptual designs have been proposed for pure laser-fusion reactors to cope with the first-wall problems typical of inertially confined fusion reactors. Three types of first-wall have been proposed: a dry, a wet, and a magnetically shielded wall. These concepts can be adapted to electron-beam and ion-beam inertial confinement systems, where the physical developments are rapid.

Other major problems of laser fusion are the development of highly efficient lasers, together with the high-reliability optical engineering that will be needed, and cheap and reliable fabrication of the target pellets. The apparently extreme demands which have to be met for the production of practical energy should not, at the present stage, deter research on laser fusion.

## Engineering assessment and future needs

No single technological obstacle which may make reactors impracticable has been found in these studies of the conceptual design of fusion reactors. But much needs to be done in order to approach realistic engineering-stage reactor design, especially in electrotechnical engineering and in establishing the practicability of materials possessing the required properties. The accomplishment of conceptual designs of D-T tokamak reactors gives rise to a much greater confidence in the outcome of fusion research.

The present designs are by no means taken as final because the development of plasma

physics, potential developments in component quality, and the introduction of new ideas will undoubtedly change the overall view of reactor systems. These designs should be regarded as revealing critical features and so stimulating further studies in improving these aspects.

The phasing of individual technological developments should be prudently chosen. Too early a detailed development of a technology might become of no use (because the general design policy changes). On the other hand, development of materials often requires a long period of time, and it is better to initiate such development early. To this category belong, among others, the super-conducting and first-wall materials. These developments may require expensive special facilities, and the provision of these on an international basis would be helpful.

## Material resources

Nuclear fusion will be based initially on the D-T reaction and will therefore consume deuterium and lithium. The supply of deuterium is practically unlimited, and the known high-grade reserves of lithium amount to about 10 million tonnes. Both $^6Li$ and $^7Li$ are consumed in tritium breeding and, although it is theoretically possible to consume all the lithium, current designs indicate that only 7–15 per cent of all lithium would be consumed.

The lithium needed in a reactor depends on blanket composition and loading; current designs indicate around 200 tonnes of natural lithium per gigawatt (electric), or $10^9$ W. Thus $10^7$ tonnes of lithium could provide the inventory for more than $10^4$ GW$_{(e)}$ of eractors, representing an energy reserve of

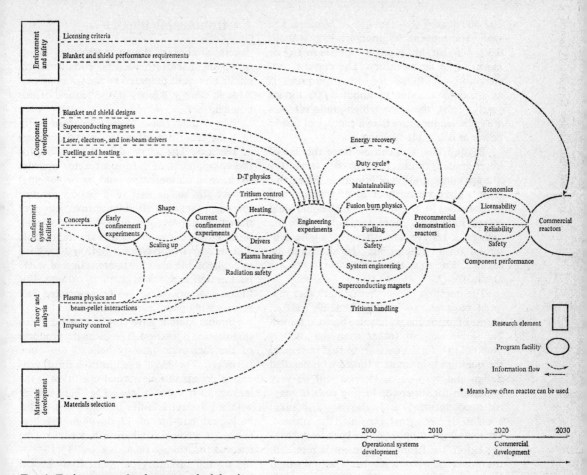

FIG. 1. Fusion power development schedule aims for commercial demonstration by the early twenty-first century. (Department of Energy, United States of America. Reprinted, with kind permission of the copyright owner, the American Chemical Society, from the 2 April 1979 issue of *Chemical and Engineering News*. Schedule based on current United States Department of Energy schedule, projected budgets, and scientific and technical developments at expected rates.)

47

200–500 quads ($Q$, where $Q=10^{21}$ joules). The world's current consumption of energy is about 0.3 $Q$ per year, but it might well rise to 1 $Q$ or more in the next century. The current consumption of lithium for other purposes is assumed to amount to about 5,000 tonnes yearly. Thus, the known high-grade reserves are sufficient for more than a century of fusion power at negligible fuel cost.

Beyond this, one can envisage the use of lower-grade lithium reserves on land (the average concentration of lithium in the earth's crust is given variously as 20–65 g/tonne) and in the oceans (0.17 g/tonne), which could yield several million $Q$. Moreover, long before the high-grade reserves are exhausted, a pure deuterium fusion reactor may be shown to be feasible. For this, the world's resources are about $10^{10}$ $Q$.

The question of whether or not other material requirements could ultimately restrict the use of fusion energy has also been examined. No single element (other than the fuel) is essential to a fusion reactor, so that in general the question is intimately linked with detailed design and with costs. Helium will be very important for super-conducting coils in magnetic confinement and also as a blanket coolant. If, as seems possible, the present, cheap well-gas supplies of He become exhausted before the fusion investment is required, the more expensive atmospheric helium would have to be used. Although no insurmountable supply barriers can be identified, these are among the constraints that reactor designers may have to face.

## Environmental impact

Several studies on the environmental impact of fusion reactors have been completed, including a report prepared by the International Atomic Energy Agency at the request of our council.

The studies have been concerned with, first, the acquisition of materials and the siting of reactors. The acquisition of materials for the components of a fusion reactor appears to have a despoliation and environmental impact similar to that of a fission reactor; there is a factor of 10–100 improvement over the situation prevailing with strip-mined coal. Siting effects of fusion reactors are comparable with those of fission systems; both systems have in common the need for the rejection of waste heat as well as a lack of chemical exhaust. For fusion reactors, there are no associated fuel reprocessing plants.

Similar studies have concentrated on routine reactor operation. The ultimate products of the fuel cycle (mostly helium) are quite harmless. The most significant hazard generated on site during normal operation is the leakage of tritium—a low-energy beta emitter with a physical half-life of 12.3 years and a biological half-life of 12 days—in the biosphere. Acceptable radiation hazards will be ensured for leakage rates to about $3 \times 10^{-4}$ of the tritium inventory per year. Radiological hazards to operating personnel are likely to compare with those pertaining to fission reactors. And although no biological hazard has been demonstrated from the stray magnetic fields of magnetically confined reactors, this aspect deserves further study.

Studies concerning reactor shut-down and accidents have yielded detailed estimates of

Christie

radioactive inventories and their associated hazard potentials. It is important to note that fusion reactors have only small quantities of fuel in the reaction zone, so that there is no possibility of dangerous nuclear excursion. Detailed risk analyses will further have to consider, however, accidental releases of radioactive materials into the environment as a result of chemical explosions, fires, release of stored magnetic energy, and natural disasters.

As to implications relating to nuclear weapons, the operation of fusion reactors does not involve the production or use of plutonium or other weapons-grade materials in any way. Safeguards by surveillance would require ensuring the total absence of fissile material and fission products on the reactor site. The development of laser fusion systems does not, of itself, result in any addition to the nuclear arsenal, although the research seems to be adjacent to some fields of knowledge used in the development of weapons.

# The way past and the way ahead

Fusion research, begun twenty-five years ago, has been done on a relatively small scale because of the size of the experimental devices and the financial support available.

Significant increases in fusion research since 1970 came notably in the United States and Japan. The programmes in the Soviet Union and Western Europe have remained almost constant in terms of the scientific personnel engaged.

Further research will require larger and more expensive experimental equipment, as well as new types of facilities, as work broadens

to include many technical problems (in addition to plasma physics, the major subject of research until now).

Much of the technology needed for fusion, such as blanket designs and coolants, radiation damage studies, shielding, remote maintenance in radiation fields and tritium processing, has already been developed extensively for fission reactors. This should ease scheduling and costs for the practical demonstration of fusion.

Similarly, much of the professional training and experience demanded by the engineering aspects of fusion is already available from the field of fission reactors.

# Time-scale, effort and cost

The development of fusion power into an economic electricity-producing source is a major scientific and technical challenge. On the path to the final goal, several intermediate goals (short of commercial demonstration) could be made proximate objectives of international fusion programmes and co-operation.

First, there is the demonstration and study of the physics of burning plasmas in ignition experiments. If the work is commenced promptly and adequately supported, such a goal could be reached by 1984–86. Next, there is the demonstration of a sustained net energy output from a fusion experiment, which then should be possible in a further seven to ten years' time. After would come development and demonstration of the technical practicability, economy and reliability of all necessary components for a fusion reactor plant, including the operation of a full demonstration plant. Last would come comparison of the economy and reliability of different reactor types. An additional intermediate goal is being pursued in the Soviet Union, namely the demonstration of breeding fissile material in a specially designed blanket by means of fusion neutrons.

The main goal is the fully operating demonstration plant, and this could be begun before full information is available from earlier stages. It should be possible to infer the commercial practicability (or otherwise) of fusion from this stage, but demonstrated commercial competitiveness would follow later.

As to cost estimates for generating electricity by fusion, the expenses for fuel and fuel processing are negligible, but the capital costs are likely to be high. Rough estimates suggest capital costs of the same order as those for the fast breeder and its associated fuel cycle. One of the main purposes of fusion research, at present, is to find ways of reducing capital costs to economically desirable levels.

Estimates of the time and cost required for a demonstration plant have been made in the United States. The times range from fifteen to forty-five years, and the cost is estimated to be $15 \times 10^9$. The time and cost for the development of fusion power depend, within limits, on national and international commitments to such a goal—not only of financial but also of intellectual resources, institutional arrangements, the degree of international co-operation and the way of the decision-making process.

## Conclusion

Estimates made thus far show that fusion power cannot solve the problems connected with the world's energy shortage which is expected before the end of the century; they show, however, that there is realistic hope that fusion power can begin to contribute to fill the growing energy gap by the beginning of the next century, provided that an adequate, worldwide effort will begin now. A substantial and increasing contribution to the production of electricity can probably be expected only after the first two decades of the next century, because the high capital investment necessary for a fusion plant will make experience of the long-term reliability of such plants highly desirable before a large-scale introduction of fusion power can be inaugurated. As this period of introduction can be shortened only by large financial risks, one has to save time when time is less expensive, that is, now—at the beginning of the development period.

Considering the significant technical progress in this field since 1970, and assuming reasonable commitment of resources and effective programme, institutional and collaborative arrangements, the net generation of fusion energy within fifteen years and the practical demonstration of an electricity-producing plant ten years later are reasonable goals for research and development in fusion.

Our council recognizes that the costs projected for fusion development are great and the times for its achievement long. But fusion, if demonstrated to be practical and economic, will be an enduring, ultimate energy resource. No further development of new, basic energy resources for the production of electricity will be required. The development costs are modest when compared with current annual expenditure on electricity production and petroleum imports.

# Solar energy

## 5

United Nations Environment Programme

*The sun is an inexhaustible source of energy, but many difficulties stand in the way of extracting useful energy from this source. This chapter reviews both passive and active systems for this extraction, underscoring the difficulties as well as the considerable potential of solar power. The difficulties centre around the high capital cost of equipment, for example, the cost of solar cells for producing electricity or of solar stills for water desalting plants and the inconvenience of storage arrangements that are necessary when peak sunshine hours do not coincide with the times of greatest need for power.*

Solar radiation is an inexhaustible source of energy. The solar flux at the outer fringes of the earth's atmosphere is 1.35 kW/m², a value known as the solar constant. As the sun's radiation passes through the earth's atmosphere the energy is depleted by absorption, scattering and reflection. This is due to the air molecules themselves, dust, and the naturally occurring gases such as ozone (which removes much of the ultraviolet), water vapour (which absorbs strongly bands in the near infra-red region) and carbon dioxide (which absorbs strongly bands in the middle infra-red). By the time it has reached the earth's surface, direct-beam solar radiation has been reduced in magnitude. The annual amount of solar radiation received at the earth's surface is about $1.2 \times 10^{17}$ W (or about $1 \times 10^{18}$ kWh), which is equivalent to more than 20,000 times the present annual consumption of energy of the whole world (UNEP, 1979). However, the extraction of all the energy provided by the sun is very difficult and only a fraction of this energy can be extracted; the efficiency of extraction depends on the location and the prevailing meteorological conditions.

For the design of systems to utilize solar energy, the most useful information available is the energy received on a horizontal surface (insolation) at the particular location. This is measured hourly at a large number of meteorological stations throughout the world and the data stored both as published records and magnetic tapes for use in computer programmes. Using these data, it is possible to calculate the insolation on collectors mounted at any angle and oriented in any direction, and the information obtained used to determine

Edited version of Chapter III, 'Solar Energy', UNEP Report, 1980.

the output of collectors used, e.g. in heat generating systems either month by month or annually (Morse, 1977). The daily average figures for insolation vary from about 9 MJ/m² for Antarctica to about 25 MJ/m² for the north-west coast of Australia, Saudi Arabia or Peru. Although it may be said that our knowledge of the character of solar radiation at the earth is satisfactory and commensurate with the state of the art of solar-energy technology, it should be noted that an extension of organized solar radiation monitoring together with an improvement in data quality control is desirable. Some international organizations, for example, the World Meteorological Organization, and regional ones are undertaking such activities.

The energy from the sun can be directly harnessed by developing passive and active solar systems, and it is in this respect that the term 'solar energy' is used in the strictest sense. However, solar radiation contributes through the interaction with natural processes to the development of secondary sources of energy that can be harnessed by suitable technologies (see Fig. 1). In the present report, 'solar energy' is used in its broadest sense to encompass all the energy sources that are derived from solar radiation.

Solar energy has been used in many areas for centuries, but mainly in a primitive way (for example, for crop drying in rural areas). Several solar devices for heating water, greenhouses, etc., have been developed and although at present the use of solar energy is still limited, extensive research and development programmes are under way in many countries to harness solar radiation efficiently for a broad number of applications. These range from heating and cooling of buildings, water heating, desalination, refrigeration, solar drying, telecommunications, irrigation, electricity generation to ovens for high-temperature materials processing.[1]

## Passive solar systems

A passive solar-energy system is one in which the thermal energy flow is by natural means; that is, by radiation, conduction, or natural convection. Passive solar offers a simple, economical, comfortable and reliable means of heating buildings. By careful siting of buildings, treatment of glazing, storage of excess heat in building mass, shading and ventilation, a building can be designed to be quite climatically adaptive, absorbing solar radiation during a winter day and storing heat till the night and yet rejecting summer sun. Passive cooling effects can also be achieved by control of radiation losses, evaporation, ventilation, and especially the use of building-mass heat capacity to average diurnal temperature swings, thus reducing afternoon overheating (Bahadori, 1978; Balcomb, 1979).

## Active solar systems

Any surface becomes heated when exposed to sunlight because solar radiation is absorbed and transformed into heat. If a heat-transfer medium—air, water or a chemical fluid—is

1. The literature on the different uses of solar energy is quite extensive and it is not the aim of this report to review such uses in detail. (For further information, see: Palz, 1978; UNIDO, 1978; ISEG, 1978, 1979; Sørensen, 1980.)

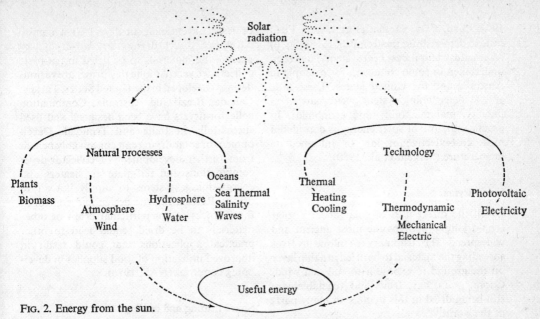

FIG. 2. Energy from the sun.

allowed to flow over this surface, it will extract heat from sunlight. Several solar collectors have been designed; these are generally classified into non-concentrating solar collectors (flat-plate collectors) and concentrating solar collectors. The former are further classified as low temperature (40–60 °C), medium temperature (60–100 °C) and high temperature (above 100 °C).[1]

## Solar water-heating

The most immediate use of solar energy is for heating water. The systems most commonly used today are solar water-heaters for residential and commercial purposes. Several countries have used solar water-heaters for many years. Japan had about 2 million units installed; in Australia, the number of solar water-heaters installed is now in the order of 70,000. Solar water-heaters are being produced in many other countries, for example, in the United States, the United Kingdom, France, the Federal Republic of Germany, Israel, Niger, Mali, Senegal, etc.; in several other countries governments are sponsoring active research and development of solar water-heating. Solar swimming-pool heating has been developed in some countries, for example, in Australia (Morse, 1977) and in the United States. Considerable potential exists for the utilization of solar energy as an alternative major resource in industrial heating generating systems (Morse, 1977; Morse et al.,

1. For a detailed description of different solar collectors see, for example, Palz (1978); Cooper (1979) and Giutronich (1979).

1977; Read, 1978; Sargent et al., 1980). Field tests to demonstrate the technical feasibility of solar industrial-process heat systems have been undertaken in some countries, for example in Australia and the United States. These tests aim at determining system performance, reliability, maintainability and economics. In practice, the use of solar energy is considered most cost-effective at low or intermediate temperature (Sargent et al., 1980).

## Solar drying

Of all the direct uses of solar energy, solar crop-drying is perhaps the most ancient and widespread. The customary technique involves spreading the material to be dried in a thin layer on the ground to expose it to sun and wind. Copra, grain, hay, fruits and vegetables are still being dried in this manner in many parts of the world.

In recent years innovations have been adopted, particularly for fruit-drying, in which fruit is placed in carefully designed racks to provide controlled exposure to solar radiation and wind, and to improve material handling. Solar drying is of special interest in the case of soft fruits, meat and fish, which are particularly vulnerable to attack by insects, as the sugar concentration increases during drying. For dried fruits and vegetables, sun drying is the cheapest and simplest method in regions having abundant sunshine, and where the post-harvest season is characterized by low relative humidity and little or no rainfall. Although there is no significant commercial manufacture of solar crop-dryers, a number of experimental designs are now in use (see, for example, UNIDO, 1978, Keener et al., 1978). These range from the use of solar-heated air in more

or less conventional air-dryers to a combination of direct drying and air-drying by placing the materials to be dried in flat-plate collector-dryers. Among the former are various designs developed in the United States, Turkey, Canada, Brazil and Australia. Combination collector-dryers have been designed and used successfully in India and Trinidad. Development of solar drying can conceivably benefit from further development of collector-dryer combinations and flat-plate air heaters and energy-storage systems to supply hot air to dryers. Research in the design and control of these processes, for particular crops or other materials to be dried, could lead to other practical applications that could result in improved utilization of food supplies in developing countries (NAS, 1976).

## Space heating and cooling

The technology of solar heating where, for example, water is the medium for heat transfer is essentially an extension of the technology employed in solar water-heating, except that heat has to be recovered from the tank through a hot-water pump. Collectors and storage units much larger than those employed in solar water-heaters are necessary to provide a substantial portion of the heat requirements. Systems employing air as the heat-transfer medium between the collectors and a storage bin containing small rocks have also been used successfully. Solar heat is stored as heat in the rocks, and is recovered when needed by passing air over the rocks and thence to the rooms. Several experimental solar houses have been built and operated with the heating system comprising a collector, a heat-storage unit, an auxiliary heater, and appropriate heat-

distribution and control systems. There has been much diversity in concept and design and some of the systems are associated with heat pumps. An extensive review of the possible methods of solar cooling for air conditioning is given by Duffie and Beckman (1974). The most intensively studied system to date is the lithium bromide/water absorption process, in which the generator receives hot water from the solar system at temperatures on the order of 80–90 °C. The development of collectors of reasonable cost that can operate efficiently at temperatures well above the boiling point of water will allow consideration of the well-known ammonia/water system.

## Solar refrigeration

Closely related to air conditioning, solar refrigeration is generally intended for food preservation or for storage of biological and medical materials. There have been experiments in several countries including the United States, Sri Lanka, France, and the USSR on solar-operated coolers using absorption cooling cycles. Most of these are aimed at household-scale food coolers or small-scale ice manufacture.

## Thermal storage

Storage is needed in all solar heating systems to avoid wasting the heat generated by the collector when this exceeds the instantaneous system load. With present technology, it is not practicable to store more than a few days' input to a system, but this is sufficient to enable the solar input to be matched to the load for periods of about a month when the average weekly load is approximately constant

(such as domestic hot-water services). Small hot-water systems use insulated copper or glass-lined steel tanks, but there are not enough large systems for a standard practice to have emerged. Copper, concrete, several salts or stainless steel have also been used. Rock storages have been used experimentally with air heaters, but, though satisfactory, are somewhat bulky. Where the product itself can be stored, e.g. hot water, it is merely a question of providing an adequately sized, insulated storage tank. Off-peak electrically heated hot-water services have for many years used this principle, and such systems are readily available commercially. It is a simple matter, therefore, to incorporate such a storage method into a solar water-heating system, it being only necessary to increase the size of the tank to provide one and a half to two days' supply, and perhaps increase the amount of insulation provided. For industrial process heating and other applications where heat rather than hot water is required, storage presents a greater problem in that the tanks become rather bulky when only the sensible heat of water changing in temperature by a few degrees is available. Nevertheless, large insulated tanks are being used satisfactorily for this purpose (Morse et al., 1977).

Salt-gradient ponds (solar ponds) have been proposed for collecting and storing thermal energy from the sun for low-temperature applications, such as space-heating, water-heating, etc. (Tabor, 1963; Tabor and Matz, 1965; Nielsen, 1975). In such ponds, the non-convective character of the dense highly saline water at the bottom leads to storage of high temperatures (as high as 95 °C) and the heat can be exploited by suitable heat-exchangers installed at adequate depth. Solar ponds have been constructed in some countries, and extensive experimental work is under way, for example, in Israel, Chile, the USSR, the United States, and other countries (Nielsen, 1975), to solve the different technological and economical problems of extracting thermal energy from such ponds (see also, Mehta, 1979).

## Solar distillation and desalination

The use of solar energy for desalting sea-water and brackish water has been demonstrated in several moderate-sized pilot plants in the United States, Chile, Brazil, the USSR, and several other countries. This century-old process consists of a shallow pool of brine from which water, slowly evaporated, condenses on the underside of the cooler glass covers and runs into troughs at the lower edges to storage. Excess brine that has not evaporated is run to waste as salt water is supplied to the basin. The idea was first applied in 1892 at Las Salinas, Chile, in a plant supplying drinking water for animals working in nitrate mining and transport. The Las Salinas plant reportedly operated for thirty years (NAS, 1976). Modern developments in solar distillation have been directed to the use of materials and designs for economical and durable constructions, to reduce the distilled water cost. Among recent developments is desalination by reverse osmosis, which makes use of thermodynamic solar-energy converters.

Solar distillation requires relatively large capital investment per unit of capacity but, in properly designed and constructed systems, a minimum of operating and maintenance costs. Product-water costs depend primarily upon still productivity, service life, capital cost of the installation, and amortization and interest rates. Productivity of a solar still is dependent on the intensity and duration of the sunlight it receives. Experience shows that a still will yield about 1,300 l/m² annually, with some variations dependent on climate and design. A typical lifetime for a still constructed of concrete, glass and other durable materials is twenty years or more; other still designs involve less durable materials that must be renewed periodically. The lack of more general use of solar stills is almost entirely the consequence of high capital investment required and the resulting high cost of water produced. There seems to be little prospect that large solar distillers will be competitive with large desalting plants supplied with conventional energy sources unless fuel prices escalate greatly. However, in situations where a community or an industry requires small quantities of water, say, less than 200,000 litres per day, the solar still may be more economical than conventional desalting plants. This is particularly true: (a) in small communities where potable water is unobtainable except at very high cost; (b) in certain industrial and commercial applications where materials must be processed in a region where all available water is brackish; or (c) for watering of livestock in areas where grazing is possible if water is supplied. It is for these moderate-volume requirements that solar stills have been built.

## Solar cookers

Different types of solar cookers have been designed since the beginning of this century. Simple solar cookers may be classified as (a) concentrating parabolic and spherical dish or through collectors where the heat at the focus of the collector directly heats either a vessel containing the food or the food itself, and (b) ovens or food warmers which are insulated boxes with transparent covers in which solar energy is collected by direct radiation or by radiation from some type of reflective surface. A solar cooker with parabolic mirror was developed and marketed in India as early as 1950. Although it was sold for a low price, it did not attract much interest. Few housewives could become accustomed to the 'new' cooking process which was not something done by generations of ancestors. Also the cloudy weather (which reduces the efficiency of cooking) and the need to warm food at times when there is no sun, added to the housewives' frustration (see Walton et al., 1978, for evaluation of solar cookers). The principal requirements for the successful use of a solar cooker in rural areas in developing countries may be summarized as follows: (a) the unit should cook food effectively; it should therefore provide energy at a sufficient rate and temperature to properly cook desired quantities and type of food; (b) it should be sturdy enough to withstand rough handling, wind and other hazards; (c) it should be sociologically acceptable and fit in with the cooking and eating habits of the people, i.e. provide for cooking to be done in the shade and if possible at times when the sun is not shining; (d) it should be capable of manufacture with local materials and by local labour; (e) it should be possible for the user to obtain a cooking unit at a sufficiently low cost. In order to overcome the problems of cooking in direct sunshine and cooking only when the sun is shining, two advanced solar cooking devices have been proposed. The first involves the use of a heat-transfer system to permit cooking to be done inside the house. The second involves the use of some type of energy storage system which would permit the cooking to be done in the evening or at other times when the sun is not shining. Such types of cookers are considerably more expensive than simple open-air cookers.

## Mechanical energy and electricity from solar energy

The term 'solar engine' designates an engine operated by solar energy. The thermodynamic cycle of such an engine may be as follows: vapour is obtained when a liquid working fluid is heated by solar radiation. This vapour expands in a reciprocating or rotating engine, doing work. From the engine it flows to a heat exchanger, in which it condenses. The condensate is reinjected by a pump (usually operated by the solar engine itself) to another heat exchanger, in which it evaporates, closing the cycle. 'Low' temperature solar engines make use of flat-plate collectors and work with temperatures below 100 °C, while 'medium' and 'high' temperature machines work with temperatures above 100 °C and in this case focusing solar collectors that track the sun are used. The most extensive application of 'solar engines' is for water pumping. Several solar-irrigation systems have been developed in many countries; solar pumps have been installed in Senegal, Upper Volta, Mauritania, Niger, Mali, Sudan, Chad, United Republic of Cameroon, Mexico, United States, Australia, etc. The first generation of these pumps are rated at 1 kW; later generations are rated at up to 100 kW (for example, in Mexico, 25 kW, 900 m³/day in 1976; in Mali, 75 kW, 9,000 m³/day in 1979), and such pumps are commercially available. The present trend in solar irrigation systems is to utilize different systems in order to increase the temperature of the working fluid and thus increase the overall efficiency of the system (Phéline, 1979).

## Solar thermal-electric conversion

Solar thermal-electric conversion systems (STECs) collect solar radiation, convert it to thermal energy first and then to mechanical energy via a thermodynamic cycle to drive an electro-mechanical energy-conversion device to generate electrical energy. STECs have the potential to convert up to 25 per cent of the incident solar radiation into electricity. There are two broad categories of STECs: (a) distributed-collector systems and (b) central-receiver systems. In distributed-collector, central-generation systems, solar energy is collected throughout the collector field and the heat is transported to a central energy-conversion plant via a pumped fluid or chemical through a piping network. In another approach, electricity is generated at each collector and brought to a central point for transmission. This concept is known as distributed collection with distributed generation. In the central-receiver system, a large field of steered mirrors (called heliostats) reflect solar radiation to a single receiver mounted on a central tower. The solar collector field, in effect, is a large parabolic steerable reflector with the focal length equal to the tower height (typically, about 100 m). With a proper design for the receiver, high-temperature steam is generated which is then used in a steam turbine driving a generator to produce electrical energy.

An active development programme on STECs is being pursued in the United States, with plans for the construction of units varying in size from 5 MWe to a 100-MWe unit to be constructed in the mid-1980s. Similar development plans for STEC systems in France, Italy and other EEC countries are under way (Phéline, 1979). A typical 100-MWe STEC unit might be envisioned as consisting of 12,500 heliostats, each having a reflecting surface of approximately 40 $m^2$, with a central receiver tower 250 m in height, supporting an absorber to provide steam or hot gas to a turbine for periods ranging from six to eight hours per day. Present designs incorporate either conventional fuel storage for operation of the plant in a hybrid mode or thermal storage sufficient to extend operation to the intermediate power-generating mode. Overall efficiencies are expected to be in the range of 15–20 per cent. Reliability of energy supply by means of STEC systems will be a function, in part, of the siting strategy chosen and the nature of the total utility system with which STEC units would be interconnected. By 1990, there should be sufficient experience with operating units in the range of 50–100 MWe so that a meaningful estimate of STEC's economic attractiveness can be made (Auer et al., 1978). It should be noted that STECs are most suited for areas with high solar radiation and low cloudiness such as those located in arid and semi-arid zones in many developing countries.

## Photovoltaic conversion

A second way of using sunlight for electricity production is by photovoltaic conversion (Palz, 1978). Photovoltaic cells, containing a thin slice of crystalline silicon or a thin film of cadmium sulphide protected from the weather by a suitable transparent material, generate direct electric current when light shines on them. The stronger the light, the more electricity is generated. With present technology, silicon cells are more efficient than those using cadmium sulphide (up to 15 per cent) but are more expensive. Solar cells are generally rated in terms of peak power: how much electrical wattage they produce in direct sunlight. The energy can be stored in batteries and used as needed. If the power is required evenly around the clock, about 5 to 10 peak watts of capacity will be required for every average watt desired. Photovoltaic cells are expensive at present and research and development are under way to increase the efficiency of the cells and reduce their costs. In 1976, the cost per peak kilowatt was about $15,000; in 1978 it was about $11,000; this year the cost dropped to about $5,000/kW and is expected to be less than about $500 in 1985 (Auer et al., 1978). This corresponds to about $60 per m² photovoltaic array at 12 per cent efficiency. The potential advantages of photovoltaic conversion are impressive. There need be no moving parts; lifetimes can (in principle) exceed 100 years, although cell performance may be expected to degrade continuously over its operating lifetime; maintenance involves little skill; both direct-beam and diffuse solar radiation can be utilized effectively; the system is inherently modular and readily lends itself to the design of virtually any system size, small or large.

At present, photovoltaic solar converters are restricted to the range of a few kilowatts, in which their use is developing rapidly. In remote or developing rural zones, extensive applications are for refrigeration (preservation of vaccines, food, etc.) water pumping and electric lighting. Some typical examples are: solar-powered navigation lights around the airport of Medina, Saudi Arabia; a navigational lighthouse in Indonesia; solar panels for battery charging, solar water pumps, educational television in Niger (Palz, 1978; NAS, 1976); telecommunications (Holderness, 1978; Phéline, 1979).

### Energy storage

Since energy storage is an essential component of most of the renewable energy systems currently under development (solar, wind, sea thermal, etc.), it is essential that storage mechanisms be developed which will operate economically at the multi-megawatt level if such energy sources are to have a significant impact on the power systems of the future. Energy can be stored in chemical form in secondary batteries, the best known of which is the lead-acid battery. In their present state of technology, lead-acid batteries cannot satisfy the requirements of either electric cars or the storage of off-peak electrical energy of utilities. Other commonly available batteries such as nickel-cadmium and silver-zinc batteries are too expensive to be used on a large scale and are severely limited by the availability of material. Advanced battery systems under development (e.g. lithium-sulphur, sodium-sulphur, aqueous zinc-chlorine, etc.) are predicted to have round-trip efficiencies of 70 to 85 per cent. Hydrogen holds promise of being a versatile and efficient energy carrier. Because of its flexibility, hydrogen is considered as the fuel of the future and hydrogen energy storage systems are being increasingly recognized and developed for a variety of applications. Electrical energy from a variety of sources (off-peak utility power, electricity generated by solar cells and wind-electric systems are some of the commonly suggested inputs) is used to dissociate water into hydrogen and oxygen gases. The hydrogen can then be stored as compressed gas or as metal hydride for use when required.

# References

AUER, P. L., et al. 1978. Unconventional Energy Resources. *World Energy Resources 1985–2020. World Energy Conference*. IPC Science and Technological Press.

BAHADORI, M. N. 1978. Passive Cooling Systems in Iranian Architecture. *Scientific American*, Vol. 238, p. 144.

BALCOMB, J. D. 1979. Passive Solar Systems. *Proceedings of Solar Energy Today Conference*. Melbourne.

COOPER, P. I. 1979. Non-concentrating Collectors. *Proceedings of Solar Energy Today Conference*. Melbourne.

DUFFIE, J. A.; BECKMAN, W. A. 1974. *Solar Energy Thermal Processes*. New York, John Wiley.

GIUTRONICH, J. E. 1979. Concentrating Collectors. *Proceedings of Solar Energy Today Conference*. Melbourne.

HOLDERNESS, A. L. 1978. Solar Power for Telecommunications Search. *Australian and New Zealand Ass. Advmt. Sci.*, Vol. 9, No. 143.

ISEG. 1978. *International Solar Energy Congress Abstracts*. New Delhi. 3 vols.
——. 1979. *International Solar Energy Congress Abstracts*. Atlanta, Ga.

KEENER, H. M., et al. 1978. *Simulation of Solar Grain Drying*. Ohio Agricultural Research Center. (Agricultural Engineering Series, 102.)

MEHTA, G. 1979. *Salt Stratified Solar Ponds*. (UNITAR Conferences on Long-term Energy Resources.) New York, United Nations. (Paper CF7/XIII/5.)

MORSE, R. N. 1977. *Solar Heating as a Major Source of Energy for Australia*. Tenth World Energy Conference, Istanbul. (Paper 4.2–3.)

MORSE, R. N., et al. 1977. *Solar Energy as Heat and for Fuel*. Tech. Conf. Inst. Engng, Canberra, July.

NAS. 1976. *Energy for Rural Development*. Washington, D.C., National Academy of Science.

NIELSEN, C. 1975. Salt-gradient Solar Ponds for Solar Energy Utilization. *Environmental Conservation*, Vol. 2, p. 289.

PALZ, W. 1978. *Solar Electricity*. London, Butterworth.

PHÉLINE, J. 1979. Énergie solaire et production décentralisée d'électricité. *Annales des mines*, April.

READ, W. R. 1978. Solar Water Heating for Domestic and Industrial Use. *Search*, Vol. 9, p. 130.

SARGENT, S. L., et al. 1980. Solar Industrial Process Heat. *Envir. Sci. Technol.*, Vol. 14, p. 518.

SØRENSEN, B. 1980. *Renewable Energy*. London, Academic Press.

TABOR, H. 1963. Large Area Solar Collectors for Power Production. *Solar Energy*, Vol. 7, p. 189.

TABOR, H.; MATZ, R. 1965. A Status Report on Solar Pond Projects. *Solar Energy*, Vol. 9, p. 177.

UNEP. 1979. *Solar-2000*. Nairobi, United Nations Environment Programme. (UNEP Report, 8.)

UNIDO, 1978. *Technology for Solar Energy Utilization*. Vienna, UNIDO. (Development and Transfer of Technology Series, 5.)

WALTON, J. D., et al. 1978. *A State of the Art. Survey of Solar-powered Irrigation Pumps, Solar Cookers, and Wood-burning Stoves for Use in Sub-Sahara Africa*. Atlanta, Ga., Georgia Institute of Technology.

# Coal and its prospects

## 6

V. Kouzminov and T. Beresovski

*World coal resources are so large that these alone could meet man's energy requirements
for at least another hundred years. Present techniques of coal production are so
inefficient that less than 10 per cent of existing resources can be used. This utilization can be
greatly increased through improved techniques of mining and of burning coal. R&D on
fluidized-bed combustion and on magnetohydrodynamic coal-fired generators holds great promise.
So does research on synthetic fuels extracted from coal. Coal must be an important alternative
to bridge us over the remaining decades of this century while new sources of energy are
being developed.*

## Overall potential

According to the World Energy Conference's
estimations, world coal resources—those that
may confidently be presumed to exist—amount
to about $10.10^{12}$ tons of coal equivalent (tec).[1]
This vast store contains twenty-five times as
much potential energy as the world's known
reserves of oil. With coal production at
the 1978 level of $3.10^9$ tec, these resources
would be enough for several thousand years.
In the event of mankind turning its back on
all resources of energy except coal, and allowing for continued increase in world energy
consumption, which during the recent years
has amounted to about 3.5 per cent, and for
unavoidable wastage, there would still be
enough coal to meet requirements for at least
another hundred years, or until new energies
such as fusion or solar towers have got past
the R&D stage and are available for wide use.
In other words, despite the fact that during the
last century, and half of the present one, coal
has been the principal source of the world's
energy, an enormous amount remains to be
exploited. At the same time, it must be pointed
out that of these resources less than 10 per cent
would be considered suitable for utilization
under present extraction techniques and economic requirements for energy sources. However, this percentage can be greatly increased
depending on the progress of R&D in mining
technologies, and it is this point that provides
the main hope for coal in the near future.

Unpublished manuscript from Unesco files,
Unesco Secretariat, Paris.

1. Tec = ton equivalent coal (unit of energy used
   in international statistics).

In fact, the period since the 1950s has seen a decline in the rate of coal extraction and utilization with the result that coal now contributes no more than 30 per cent to the world energy balance, while oil and natural gas have taken over its former dominant role. Forecasts indicate that this rate of coal production and utilization will be maintained during the next forty to fifty years, but its absolute amount will be significantly increased. The World Energy Conference analysis, as well as estimations of other organizations and individual experts, indicate that by the year 2000 coal production should reach $5.8 \times 10^9$ tec (about twice the present level) and by 2020 it will be about $9 \times 10^9$ tec. Such an enormous increase of coal production will only be possible with the development and adoption of new approaches to the technology of coal production and utilization.

The major constraints preventing a more rapid increase in coal production and utilization can be listed as follows:

Mining investments are highly capital-intensive and coal-mining is still a costly process with risks for human health and the environment.

Present methods of coal transportation from pit-head to power-station are labour-consuming and inefficient.

Burning coal emits harmful gases that are hard to eliminate, as well as particles of gritty matter, and leaves large quantities of slag.

Coal cannot be used in its natural state as fuel for most types of engines.

Additional environmental problems which should be mentioned are: first, the issue of carbon dioxide ($CO_2$) built up in upper reaches of the atmosphere, the increasing levels of which could cause changes in the global climate. This is being studied in many countries and limitations could be introduced to the utilization of fossil fuels including coal.

Second, emission of $SO_2$ and $NO_x$ as a result of coal combustion has led to so-called acid pollution, i.e. rainfalls containing acids at long distances from emission plants. While new methods of pollution control are being developed, this problem is likely to require increasing attention in the future.

The major developments for the future which promise to overbalance these constraints lie in three main directions: advances in current technology for production of coal, improvements in the use of coal and the application of new technologies for the production and use of coal. Active research programmes are under way to deal with problems on each of these levels and in many cases technological developments already exist or are in an advanced stage of testing.

## Advances in current technology for coal

A partial solution to the problems of investments and mining cost is the development of open-cast mining wherever feasible, which is considerably less capital intensive with lower production costs. However, there are significant factors limiting the development of open-cast mining: the availability of reserves appropriate for this method of coal extraction, and environmental problems dealing with the destruction of natural landscapes and the need for recultivation of dumps. The large-scale mechanization and automation of mining processes possible in such open-cast mines make extraction a much less expensive business and reduce the health hazard to mine-workers. Similar advances are also possible in underground mines where they can lead to significant economies in production.

One of the most promising methods of coal extraction, which allow considerable reductions in capital investments and mining cost, is underground gasification (which is discussed further below). Much promising research is being carried out on a number of approaches that will change the nature of the coal industry as it now exists.

In recent years, progress has been made in dealing with transportation problems, which are the most serious in the case of new coal regions located as a rule at remote distances from consumers. At present, the possibility of creating complex transportation infrastructures for coal gains more and more supporters. Economic analyses show that the cost of coal transportation depends on the type used and the distance carried. For example, transport by barge was found to be the least costly for carrying coal distances of more than 160 km. However, barges are limited to areas where waterways exist. Railways are more accessible than barges and the distances saved by rail may reduce overall transportation costs, but maintenance costs drive up the price of rail transport.

Pipelines are alternative means of long-distance coal transportation, but have not been extensively used to date. Pneumatic pipelines use air pressure to move coal, much as a vacuum cleaner does, and in general terms they could be the most effective means of transporting coal for distances of less than 24 km. Quite recently, some slurry pipelines for transporting coal over longer distances have begun operation and have

showed their high effectiveness. To make coal transportation more effective, the transport infrastructure might best involve a combination of the above means where they are economically justified.

Another approach to the coal-transportation problem has led to the creation of mining/electric-power complexes, from which electric power is delivered to consumers by high-tension transmission lines over long distances. This approach is especially suitable for countries with large brown-coal deposits, i.e. coal with low caloric value. The construction of mining/power/chemical complexes producing synthetic oil and gas and other chemical products at the pit-head could also contribute to the solution of coal-transport problems.

## Improved use of coal

Another and probably most important area to be considered is the improvement of the technology of coal use. One promising method is fluidized-bed combustion of coal. This method is based on a new boiler technique where coal is burned as a bed of particles maintained in a state of fluidization by a flow of air which also provides the oxygen for the combustion process. In a fluidized bed, a uniform temperature in the boiler permits efficient combustion at medium temperatures (800–900 °C), so there is no melting of ashes and subsequent deposition on boiler surfaces. Moreover, it is possible to use non-corroding gases, resulting from the combustion, in a gas turbine to generate electricity.

Very high rates of heat-transfer can be achieved within the fluidized bed so that there is an advantage to introducing boiler tubes into the bed with a subsequent reduction in the length of boiler tubes. Calculations and experiments show high thermal efficiency of fluidized-bed combustion of coal varying from 78 to 95 per cent. In some pilot plants, the thermal efficiency has reached 99 per cent, with relatively low emission of polluting gases. Furthermore, the use of limestone or other suitable substances in fluidized beds permits the capture and removal of $SO_2$ during the combustion process within the bed; this is an important advantage for coal with a high sulphur content. Thus, from the environmental point of view, fluidized-bed combustion is one of the most promising methods of coal utilization currently under development.

Estimations based on pilot-plant experience indicate that capital costs of plants using fluidized-bed combustion of coal could be 15–25 per cent lower than those of conventional plants. The overall efficiency of a thermal power-plant with this technology could reach a level of 45–50 per cent, which is about 10 per cent higher than that of the present plants.

In recent years, progress has been made in fluidized-bed combustion R&D. Such countries as the Federal Republic of Germany, the United States, USSR, Sweden, Italy and Czechoslovakia have research plants and furnaces utilizing this method of coal combustion with heating capacities up to 60 MW. The design of a 200-MW plant to be constructed by 1987 is under way in the United States.

Another approach to improve the efficiency of thermal power-plants consuming coal may be through the introduction of magnetohydrodynamic (MHD) coal-fired generators. In an MHD generator the expansion of an electricity-conducting working fluid through a magnetic field produces electrical energy directly through the electromagnetic interaction between the moving fluid and the magnetic field. The electric current produced in MHD generators is captured by pairs of electrodes distributed along the generator channel. It is important to stress that MHD is a direct method of converting thermal energy into electricity. The magnitude of the electric current depends on the electrical conductivity of the working fluid or gas. If the gas is at very high temperatures, it is ionized and becomes electrically conductive or in a plasma state. If it is below 2,000 °C (normal temperature of coal combustion) some 'seed' material (usually easily ionized alkaline metal compounds) must be injected into a combustion chamber in order to increase its conductivity, which is essential for the efficient operation of an MHD generator. After being used in the MHD generator, the combustion gases are still sufficiently hot to be utilized for the conventional steam cycle. Therefore, the coal-fired MHD generator is considered as a high-temperature 'topping' cycle for steam-turbine power plants. High temperature co-generation of this nature could greatly increase the overall efficiency of an installation perhaps up to 60 per cent.

Another very important advantage of the coal-fired MHD generators is a considerably reduced emission of dangerous gases. The need to remove and recover alkaline 'seed' material guarantees the removal of heavy particles from stack gases, since electric precipitators are used at the exhaust end. On the other hand, the seed material can act as a natural absorber of sulphur and can be reconverted into the carbonide required in the initial seeding process and into pure sulphur which can be used as a by-product. Techniques also exist to deal with the problem of nitrogen oxides, and it can be said that the potential of MHD systems to meet pollution-control requirements can be an important point in their development.

The most advanced MHD technology now exists in the USSR and the United States, but the best results to date have been achieved in MHD generators utilizing natural gas. The largest MHD electrical power-plant of 500 MW is now under construction in the Soviet Union. Quite recently the development of coal-fired MHD generators has been started in the USSR and in the United States. In the near future, these two countries are considering the construction of a few coal-fired MHD plants with electrical capacities of 20–50 MW. International scientific co-operation is playing a role in promoting MHD R&D. It is hoped that the first experts' meeting on coal-fired MHD, held in Sydney (Australia) in November 1981 under Unesco's sponsorship, will contribute to the achievement of this technique of coal utilization.

## New technologies

Synthetic fuels from coal could open a new chapter in the history of coal and could be substituted for natural oil and gas in engines. Theoretically the chemical processes of obtaining synthetic oil and gas are well known and now the main purpose in this particular field is to make progress in the development of applied industrial technology. Basically, conversion of coal into liquid or gaseous fuels involves the addition of hydrogen to the carbon in coal under specific high-temperature conditions sometimes in the presence of a catalyst to ensure a more efficient chemical reaction.

There are two ways to obtain gaseous fuels from coal. One is the reaction of coal and steam under high temperature and pressure in the gasifier. In this case additional energy is provided through the partial oxidization of carbon. The product of this process is called low-BTU gas with a heat value of 4–9 MJ/m³, and it can be used economically on the spot for certain applications. If in the process pure oxygen is used instead of atmospheric air, it will result in the production of a synthetic gas with a higher heating value of 9–20 MJ/m³. It is also possible to produce high-BTU gas with a heating value close to that of natural gas, which involves a catalytic reaction called methanation. The other involves the direct reaction between coal and hydrogen (hydrogenation), resulting in the production of high-BTU gas.

The underground gasification of coal, as we mentioned above, is one of the cheapest processes of coal extraction. In comparison with the gasification of extracted coal, it does not require special industrial facilities for producing synthetic gas, and this fact makes this method economically favourable. Underground coal gasification is especially applicable to coal deposits lying too deep to be economically recoverable by ordinary means. The gasification is usually carried out either with injected air under pressure so that the product is low-BTU gas, or with the injection of an oxygen/steam mixture in order to obtain synthetic gas with a higher heating value. Sometimes, hydrogen can also be injected for producing high-BTU gas.

Liquid fuels in the form of synthetic crude oil can be obtained from coal via three principal methods. In the first, synthetic gas obtained by gasification can be converted by a synthesis process into liquid fuel (the Fisher-

Tropsch process). In the second, pyrolysis (heating without oxygen) of coal allows oil to be produced, which is then treated by hydrogen to obtain high-quality liquid fuel. Finally, coal can be dissolved in a (convenient) solvent which with subsequent filtration of ashes, removal of solvent and hydrogen treatment results in synthetic oil.

The major problems in the production of synthetic gases and liquid fuels from coal are of an economic nature and there is need for a demonstration of reliable large-scale operations without harmful environmental impact. The production cost of these fuels is still considerably higher than that of natural gas and oil, and this fact is a major obstacle to the wide utilization of these methods. But there is a strong belief that these problems will be solved in the near future, since more and more efforts are being undertaken at national and international levels to improve the technology and to make the production of synthetic fuels from coal more economic. Thus the prospects are good for the eventual use of coal as a liquid fuel for transportation purposes.

More and more R&D efforts are being undertaken in the field of coal liquefaction and gasification. Large-scale research programmes in this field are now being carried out in the Federal Republic of Germany, the United Kingdom, the United States, the Soviet Union and in other countries. In some of them there are industrial facilities for producing synthetic oil and gas.

Underground gasification is in a more advanced state of development than aboveground gasification technology, and considerable progress in this field has been made in Belgium, the Federal Republic of Germany, the United States and the USSR. The Soviet Union has three industrial coal-gasification plants under operation producing $1.5 \times 10^9$ m³ low-BTU gas per year.

## Conclusion

It may be expected that the above-mentioned new coal-utilization technologies will permit more efficient and more extensive use of coal and that the role of coal as one of the principal energy sources of our planet is likely to continue and has the potential. Thus coal must continue to be an important element in any consideration of alternative sources of fuel to meet growing world needs. This will be particularly true over the remaining decades of this century while technology for other promising sources is being developed and tested prior to large-scale application.

# The soft options

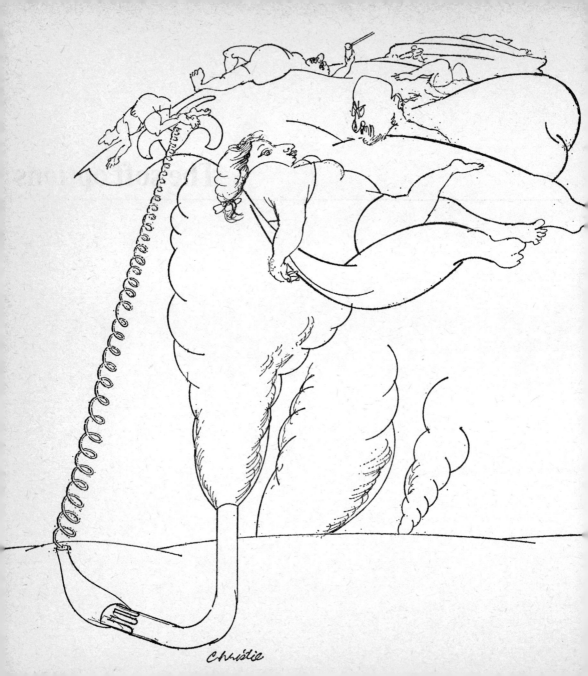

Christie

# Geothermal energy

## 7

United Nations Environment Programme

*Natural heat from within the earth can be tapped as an energy source in three ways: as hot water or steam systems, as hot dry-rock systems and as conduction-dominated systems. The first two are being increasingly exploited to generate electricity while the third lends itself to a variety of uses: heating of buildings; agriculture, especially in greenhouses; and industry. Estimates see geothermal energy providing about 0.8 per cent of total demand by the year 2000.*

Geothermal energy is based on the natural heat of the earth. The upper crust has a mean temperature gradient of 20–30 °C/km depth, and White (1965) estimated the heat stored, beyond surface temperatures, in the outer 10 km of the earth's crust to be about $12.6 \times 10^{26}$ J. This resource base is equivalent to the heat content of $4.6 \times 10^{16}$ tonnes of coal (assuming heat content of coal to be $27.6 \times 10^9$ J/tonne), or more than 70,000 times the heat content of the technically and economically recoverable coal reserves in the world.[1] However, geothermal heat in the outer 10 km of the earth's crust is too diffuse to be an exploitable energy resource on a world-wide basis. Resources suitable for commercial exploitation may be defined as localized geologic deposits of heat concentrated at attainable depths, in confined volumes, and at temperatures sufficient for electrical- or thermal-energy utilization.

## Types of geothermal resources

From the geological point of view, geothermal resources can be classified into hydrothermal convection systems, hot igneous systems and conduction-dominated systems.

### Hydrothermal conversion

Subsurface reservoirs of steam or hot water, which may display such surface characteristics as boiling springs, sulphurous mud-flats and fumaroles, are categorized as hydrothermal convection systems. The creation of such sys-

1. The coal reserves recoverable under current economic and technological conditions are estimated to be $6.3 \times 10^{11}$ tonnes (Peters et al., 1978).

Edited version of Chapter II: 'Geothermal Energy', UNEP Report, 1980.

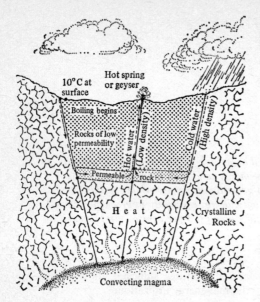

10°C at surface

Hot spring or geyser

Boiling begins

Rocks of low permeability

Hot water (Low density)

Cold water (High density)

Permeable rock

H e a t

Crystalline Rocks

Convecting magma

Fig. 1. Schematic diagram of a geothermal reservoir.

tems begins with a source of heat, hot or molten rock, that lies relatively close to the earth's surface. Overlying this high-temperature rock zone is a permeable rock formation containing water, largely of meteoric origin, which rises up as it is heated by the hot rock below (Fig. 1). Above the permeable rock is a cap of impermeable rock, which traps the super-heated water, but cracks or fissures in it allow the fluid to rise to the earth's surface either as steam (a vapour-dominated or dry-steam system) or hot water (a liquid-dominated system). Hydrothermal convection systems are usually associated with the earth crustal tectonic plate boundaries and related volcanic activity.

Hydrothermal convection systems can be subdivided into hot-water-dominated or vapour-dominated systems, depending on whether the near-surface permeable zones produce principally water or steam when tapped by production wells. Liquid-dominated systems (or hot-water systems), which are far more common than vapour-dominated ones, vary greatly in their chemical and thermal characteristics by site. Based on the temperature of the liquid, these systems can be classified into: (a) high-temperature: greater than 150 °C; (b) medium-temperature: 90–150 °C; and (c) low-temperature: less than 90 °C. High-temperature systems may be further divided by the characteristics that affect their performance (salinity, chemical composition, stratigraphic set-up, permeability of reservoir, etc.). These characteristics are important in determining the commercial viability of a hot-water system, either for electricity generation or for other purposes.

Generally, the flashed-steam process is used for electricity generation from hot-water systems. In this process, as the hot water (which is under high pressure) is withdrawn from the reservoir by wells and nears the surface, the pressure decreases, causing about 20 per cent of the fluid to boil and 'flash' into steam. Separators separate the steam from the water; the former is then directed to the turbines. The water leaving the separators is available for further processing depending upon its mineral content. The water may be reinjected into the local rock formation or may be desalinated for further use prior to reinjection (minerals may be economically recoverable). Examples of hot-water geothermal fields are Wairakei and Broadlands in New Zealand, Cerro Prieto in Mexico, Salton Sea in California, Otake in Japan.

Another method for the production of

electricity from high- or medium-temperature geothermal waters is the binary-cycle process. In this process, the water withdrawn from the reservoir is used to heat a second fluid (freon or isobutane) having a low boiling point. The vapour thus generated by boiling the second fluid is used to drive the turbine. Once used, the vapour is condensed and recirculated through the heat exchanger in a closed system, where it may be heated and used again. Equipment using freon as a second fluid is now commercially available in the range of temperatures between 75–150 °C and for capacities of 10–100 kW(e). These can be used for production of electricity at suitable sites, especially in remote rural areas (Phéline, 1979).

Vapour-dominated systems (also known as dry-steam systems) produce superheated steam (about 240 °C) at high pressures (up to 35 kg/cm²) with minor amounts of other gases but little or no water. Since the steam is usually of high quality, that is, it contains few particulates or other substances that must be extracted before use, it can be fed directly into a conventional steam turbine to generate electricity. Upon issuing from the turbine, the steam is directed through a condenser where it is condensed; a part of that liquid may be used as a coolant for the condenser and the remainder may be reinjected into the rock formation. Examples of dry-steam geothermal fields are: Larderello, Italy; The Geysers, California, and Matsukawa, Japan.

## Hot igneous systems

The second type of geothermal resource (hot igneous systems) includes both magma and hot impermeable dry rock (the solidified margins around magma and the overlying roof rock).

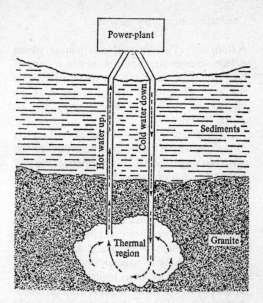

Fig. 2. Schematic diagram of a hot-dry-rock heat-recovery process.

The recovery of geothermal energy directly from magma is not yet feasible. The technology required to utilize hot dry rock is beginning to be developed. Preliminary engineering approaches to tapping the energy potential envisage a circulatory fluid flow loop through the rock (Fig. 2). First a well would be drilled into the hot formation; then cold water would be injected under high pressure to fracture the formation, and a second well would be drilled to intersect the fractured zone. Finally, cool surface water would be injected to the first well, passed over the hot dry rock, and withdrawn through the second well in the form of steam or hot water. The heated fluids generated could then be processed using either the flashed-steam or binary-cycle process.

## Conduction-dominated systems

A third type of geothermal system arises where a deep sedimentary basin occurs in a zone of high heat-flow. In situations such as the Paris Basin or Hungarian Basin, water may be tapped by wells at temperatures up to about 100 °C.

A special category of the latter type occurs in regions where the normal heat-flow of the earth is trapped by insulating impermeable clay beds in a rapidly subsiding geosyncline or downward bend of the earth's crust. Geo-pressured deposits are hotter than normally pressured deposits because upward loss of the included water has been stopped for millions of years. The high temperatures and pressures have resulted in a natural cracking of the petroleum hydrocarbons in the sediment and in geopressured zones, reservoir fluids commonly contain 1–2 m³ of natural gas per cubic metre of fluid. These dissolved hydrocarbon gases would be a valuable by-product of fluid production. Temperatures of produced water from geopressured geothermal systems would range from 150–180 °C; well-head pressures would range from 280–560 kg/cm². Production rates could be several million cubic metres of fluids per day per well, and perhaps about 30,000 m³ of natural gas per day per well (Hickel, 1972). Geopressured reservoirs have been found in many regions while searching for oil and gas; for example in the United States, Mexico, South America, the Far East, the Middle East, Africa, Europe and the USSR. The exploitation of such reservoirs for energy has not yet been demonstrated.

# Use of geothermal energy

The utilization of geothermal energy for the production of electricity and the supply of domestic and industrial heat dates from the early years of the twentieth century. For fifty years the generation of electricity from geo-thermal energy was confined to Italy (the first experimental electric-power generator was op-erated at Larderello in 1904 and power has been produced since 1913) and interest in this technology was slow to spread elsewhere. In 1943 the use of geothermal hot water for space heating was pioneered in Iceland. More recently intensive exploration work was under-taken in New Zealand, Japan, the United States and some other countries, which led to the commissioning of a number of geothermal power stations in these countries. Table 1 summarizes the world geothermal electrical generating capacity by the mid-1970s. It should be noted that geothermal fields are at various stages of exploration and development in Chile (El Tatio), Nicaragua (Momotombo), Turkey (Kizildere, Afyon), Kenya (Olkaria), Ethiopia, Guadalope, western United States and India. In addition, geothermal exploration is being carried out or considered in more than fifty countries.

The non-electrical applications of geo-thermal energy are widespread. Early uses included medicinal mineral baths which are well documented. The use of geothermal energy for space heating has become widespread in many countries; the most significant district heating system is the well-known system at Reykjavik in Iceland. Other countries using geothermal energy for space heating include the United States, the USSR, New Zealand, Japan, Hungary, France (Howard, 1975; Ban-

TABLE 1. World geothermal electrical generating capacity (MWe), 1979

| Country | Field | Installed | Planned |
|---|---|---|---|
| United States | The Geysers, Cal.[1] | 660 | 460 |
| | Imperial Valley, Cal. | — | 150 |
| | Roosevelt, Utah | 1 | 50 |
| Italy | Larderello[1] | 381 | — |
| | Travale[1] | 15 | — |
| | Monte Amiata | 25 | — |
| New Zealand | Wairakei | 192 | — |
| | Kawerau | 10 | — |
| | Broadlands | — | 100 |
| Japan | Matsukawa[1] | 22 | — |
| | Otake | 11 | — |
| | Onuma | 10 | — |
| | Onikobe[1] | 25 | — |
| | Hatchobaru | 50 | — |
| | Kakkonda | 50 | — |
| | Nigorikawa | — | 50 |
| Mexico | Pathé | 3.5 | — |
| | Cerro Prieto | 150 | — |
| El Salvador | Achuachapán | 60 | 60 |
| Iceland | Námafjall | 3 | — |
| | Krafla | — | 55 |
| Philippines | Tiwi | 110 | 110 |
| | Macban | 110 | 110 |
| USSR | Pauzhetsk | 5 | 7 |
| | Paratunka | 0.7 | — |
| Turkey | Kizildere | 0.5 | 10 |
| Chile | El Tatio | — | 15 |
| Indonesia | Kawah Kamojang[1] | — | 90 |
| TOTAL | | 1 894.7 | 1 267 |

1. Dry steam fields; the rest are hot-water dominated fields.

*Source:* Muffler (1975); Ellis (1975); UN Committee on Natural Resources, E/C7 164 (1977); Ellis (1978). For Indonesia: private communication from A. Arismunandar, 1980.

well, 1975; El-Hinnawi, 1977). In the latter, some 25,000 to 30,000 apartments near Paris are heated with geothermal energy and there are plans to extend this system to cover some 500,000 apartments by 1985–90. Hot geothermal waters are also used extensively in agriculture, especially in greenhouses, in connection with animal husbandry, in aquaculture and in controlled growth of single-cell proteins. Agricultural applications of geothermal energy currently constitute the major part of the non-electrical applications. Over 90 per cent of this capacity is associated with large acreages of greenhouses in the USSR (about 25 million m$^2$ of greenhouses are operated with geothermal waters producing about a million tonnes of vegetables per year). There is also a wide range of use of geothermal energy in industry: from the drying of fish, earth and timber, to pulp and paper processing. The two largest present industrial applications are a diatomaceous earth plant in Iceland and a pulp-paper and

TABLE 2. Some major non-electrical developments of geothermal energy

| Country | Area | Fluid temperature (°C) | Approx. harnessed energy (MWt) | Use |
|---------|------|------------------------|-------------------------------|-----|
| France | Paris Basin | 60–80[1] | 30 | Heating |
| Hungary | Szentes-Szegred | 80–90[1] | 350 | Heating, agriculture |
| Iceland | Reykjavik | 80–130[1] | 350 | Heating |
| | Hveragerdi | 180[1,2] | 20 | Heating and hot houses |
| | Námafjall | 185[2] | 10 | Diatomite drying |
| Japan | Okawa | 70[1] | 2 | Heating |
| | Various areas | 70–100[1] | 5 | Horticulture |
| New Zealand | Kawerau | 195 and 165[2] | 125 | Kraft paper-mill |
| | Rotorua | 100–175[1,2] | 50 | Heating, air-conditioning |
| United States | Boise, Idaho | 77[1] | 10 | Heating |
| | Klamath Falls, Oregon | 40–110[1] | 6 | Heating, greenhouses |
| USSR | Makhachkala | 60–70[1] | 25 | Heating |
| | Zugdidi | 80–100[1] | 60 | Heating |
| | Cherkessk | 80–100[1] | 25 | Heating |
| | Various areas | 60–100[1] | 500 | Greenhouses |

1. Water.
2. Steam.
*Source:* Howard (1975).

wood processing plant in New Zealand. It should be emphasized that there is a wide range of uses of geothermal fluids for non-electrical purposes, and that several schemes are being developed for combined production of power and process heat. Table 2 summarizes the major non-electrical uses of geothermal energy.

Different estimations have been made for the capacity of geothermal energy most likely to be harnessed by the year 2000. Table 3 gives a considered estimate of geothermal capacity likely to be harnessed by 1985 and the year 2000; the geological potential of favourable areas is given for comparison. Assuming that the total primary energy demand by the year 2000 is $600 \times 10^{18}$ J, geothermal energy that can be harnessed by that time can provide 0.8 per cent of this demand.

TABLE 3. Geothermal energy scenario until the year 2000

|  | 1979 | 1985 | 2000 |
| --- | --- | --- | --- |
| Geothermal electricity estimated from collected data [MW(e)][1] | 1 895 | 5 000 | 50 000 |
| Geothermal non-electrical energy estimated from collected data [MW(t)][2] | ~ 2 000 | 10 000 | 100 000 |

1. Geothermal electricity calculated according to geological potential is about $5 \times 10^5$ MW(e); see Auer et al., 1978.
2. Geothermal non-electrical energy calculated according to geological potential is about $8 \times 10$ MW(t); see Auer et al., 1978.

# References

AUER, P. L., et al. 1978. Unconventional Energy Resources. *World Energy Resources 1985–2020. World Energy Conference*. IPC Science and Technology Press.

BANWELL, C. J. 1975. Geothermal Energy and its Uses: Technical, Economic, Environmental and Legal Aspects. *Proceedings of the Second United Nations Symposium on the Use of Geothermal Resources, San Francisco*. Vol. 3, pp. 2257–67.

EL-HINNAWI, E. E. 1977. *Energy, Environment and Development*. Tenth World Energy Conference, Istanbul. (Paper 4.8–1.)

ELLIS, A. J. 1975. Some Geochemical Problems in the Utilization of Geothermal Waters. *Proceedings of the Grenoble Symposium of the International Association of Hydrological Science*, pp. 100–9. (Publication No. 119.)

———. 1978. *Environmental Impact of Geothermal Development*. Report prepared for UNEP.

HICKEL, W. J. 1972. *Geothermal Energy*. University of Alaska.

HOWARD, J. H. 1975. Principal Conclusions of CCMS Non-electrical Applications Project. *Proceedings of the Second United Nations Symposium on the Use of Geothermal Resources, San Francisco*. Vol. 3, pp. 2127–41.

MUFFLER, L. J. P. 1975. Present Status of Resources Development. *Proceedings of the Second United Nations Symposium on the Development and Use of Geothermal Resources, San Francisco*. Vol. 1, p. xxxiii.

PETERS, W., et al. 1978. An Appraisal of World Coal Resources. *World Energy Resources 1985–2020. World Energy Conference*. IPC Science and Technology Press.

PHÉLINE, J. 1979. Énergie solaire et production décentralisée d'électricité. *Annales des mines*, April.

WHITE, D. E. 1965. Geothermal Energy. *US Geological Survey Circular*, No. 519.

# Ocean energy on parade

Walter Schmitt

*A renewable source of energy likely to increase as part of the energy mix is the ocean: tides, waves, temperature and salinity gradients, currents and winds. Quantitative measures of energy refer both to its amount and also to its time rate of production or use (power). Moreover, the intensity or what might be called the energy density of a renewable energy source is also important since this determines the ease with which the source can be tapped. Unfortunately, ocean sources have low energy densities, necessitating very large size in the devices used for harnessing them.*

*An economic analysis of the cost of tapping ocean energy sources considers corrosion factors, capital costs, and environmental impacts. In so far as conservation appears unavoidable in bridging world energy needs, ocean sources appear useful particularly in providing possibilities of good end-use matching and load management.*

In the future, energy will be supplied from a far greater mixture of sources than is now the case. This mix will have strong geographical and economic components that vary according to proximity to resources, scale of demand and level of technological development. The energy mixture will contain, in all likelihood, a number of renewable sources of energy available from the oceans. The fuller spectrum of ocean energies would include tides, waves, temperature and salinity gradients, currents and winds.

Table 1 lines up all the accessible forms of ocean-associated energies and classifies them according to origin, lifespan and best use.

The lines of distinction are slightly blurred, however, with the inclusion of winds and the exclusion of biomass. Both can be acquired on land as well as in or on the ocean, but the large marine plants considered for biomass harvest play an important role in the sea's food resources while biomass of land plants is a thousand times greater resource.

There is an important distinction I wish to make here with respect to the use, specifically the rate of use, between renewable, quasi-renewable and non-renewable energy forms. If we deal with a renewable form, say waves or thermal gradients, we can at most intercept all the wave flux or all the temperature-

From *Harvesting Ocean Energy*, p. 17, Paris, Unesco, 1981.

TABLE 1. Ocean-associated energy resources

| Origin | Form/source | Lifespan | Realm | Best Use |
|---|---|---|---|---|
| Genesis | H, D, Li (fusion) | Non-renewable[1] | Sea-water | Thermal, process steam |
| Genesis | $U^{235}$, $Th^{232}$, $K^{40}$ (fission) | Non-renewable[2] | Sea-water Crust | Thermal, process steam |
| Geological accumulation | Oil, gas, coal | Non-renewable | Sediments | Petrochemicals, thermal |
| Compressive and radioactive heating | Geothermal | Quasi-renewable | Crust Springs | Thermal |
| Earth–moon rotation | Tides | Quasi-renewable | Sea-water | Kinetic, electrical |
| Thermo-nuclear fusion in sun | Waves | Renewable | Sea-water | Kinetic, electrical |
| | Temperature | Renewable | Sea-water | Thermal, electrical |
| | Salinity | Renewable | Sea-water Salt deposits | Chemical, electrical |
| | Currents | Renewable | Sea-water | Kinetic, electrical |
| | Winds | Renewable | Ocean surface Land surface | Kinetic, electrical |
| | Biomass | Renewable | Sea-water Land surface | Organic substrate |

1. Strictly speaking non-renewable, but resource sufficiently large to last for many millennia.
2. Strictly speaking non-renewable, but resource sufficiently large to last for many centuries.

gradient flux. That is the outside natural limit for that use, and one would prudently extract only a small fraction of the flux, lest biological processes that depend on it suffer significant changes.

A non-renewable form, for instance coal, could however be extracted and used in one year or in a thousand years. Technically the limit would be set by man: what is not used now can be used later; this is not so with renewable energy. But considering pollution and environmental impact, the use of coal should proceed at a self-imposed, small rate.

The quasi-renewable forms of energy, tides and geothermal, are also subject to environ-

mental constraints if fully exploited. There are, moreover, only a relatively small number of favourable sites where these two forms are intense and technically accessible.

Perhaps the distinction is not so important after all. In each case a prudent, moderate rate of use, consanguine with the health of the biosphere, mankind's womb and matrix, is called for.

## The resources

While the terms 'energy' and 'power' are often used interchangeably, in physics and engineering their meaning is different but related. Power is simply the rate at which energy is generated or used. Energy is measured in calories, BTUs, joules, kilowatt-hours or foot-pounds, and power in calories, BTUs per minute, kilowatts or horsepower.

To illustrate the difference, visualize a hydroelectric reservoir. The stored water represents energy; it can be expressed in kilowatt-hours and can be used slowly or quickly, depending on the size of the generator. Thus, a one-million-kilowatt-hour (energy) reservoir could be emptied in 10,000 hours by a 100-kW (power) generator or in 10 hours by a 100-MW (power) generator. In other words, the reservoir could supply electricity to a small village for a year or to a small city for half a day.

Power is the correct term for describing the flux of renewable energy in nature. (Oil, coal or uranium reserves, however, are stated in terms of stored energy.) In Figure 1 are shown: energy fluxes in nature (top line), technically feasible power potentials (middle line), and practical power potentials (bottom line).

In sequence, the natural energy fluxes describe: currents over the entire ocean, power dissipated by all tides including rock tides, integration in space and time of waves, mixing of (fresh) precipitation into (saline) sea-water, the 20 °C temperature difference between surface- and deep-water in the tropical oceans, and the ocean's share of the global, near-surface winds.

Technical feasibility means: restricting currents to major currents; tides to high ranges; waves to energetic regimes; salinity to the large run-off to sea-water gradients; temperature to the Carnot-cycle efficiency for 20 °C; and winds to the theoretical limit of 59 per cent.

Practicality means: limiting current slow-

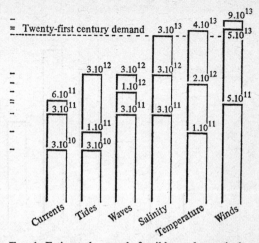

FIG. 1. Estimated natural, feasible, and practical power potentials (in watts). (Based on data from Wick and Schmitt (1977) and Gustavson (1979).)

down to one per cent; applying a 30 per cent plant factor to tides; intercepting waves only near-shore at 30 per cent efficiency; discounting salinity to 10 per cent for unknown process efficiencies and siting constraints; assigning a 50 per cent plant factor and 10 per cent siting constraint to temperature; and assuming 1 per cent for optimal wind conditions and proximity to shore.

The power output is in each case assumed to be electricity. For comparison the present world installed electrical generating capacity is about $10^{12}$ watts, or one terawatt (TW), with roughly one-third in the United States, one-third in Europe, and one-third elsewhere. In order to allow for population growth and more equitable distribution, a demand of $3 \times 10^{13}$ watts is sometimes projected for the twenty-first century (Isaacs and Seymour, 1973).

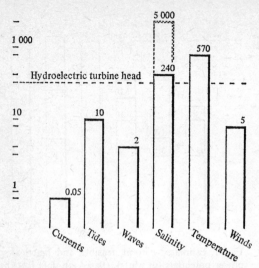

FIG. 2. Energy density, in equivalent metres of water head. (Wick and Schmitt, 1977.)

Besides potentials, there is another important aspect of the use of renewable energy sources; that is, their intensity or energy density. It is easiest technically to utilize sources with the highest energy density, since conventional, high-pressure, compact technology can be employed. As is evident from Figure 2, these ocean sources have low energy densities or heads as compared with hydroelectric power. Salinity and temperature-gradient heads are not strictly comparable; as chemical and thermodynamic energies they are more difficult to utilize than kinetic (mechanical) energies. Much of the prospect for bringing ocean energies into use consequently depends on developing low-pressure technologies, as has been done successfully for tidal power. This means a need for large components, such as the giant rotors in tide, current and wind

turbines, or the passing of large volumes of sea-water along extensive heat exchangers or salinity membranes.

## The economics constraint

This discussion provides a good introduction to the economics of ocean-energy extraction. Adaptation of conventional, high-pressure technology will be less costly the higher the energy density, but the costs will also depend on the scale and location of the plant, and the status of the local economy. In some places, alternative sources of energy are already competitive.

The sea is typically a more expensive place in which to build and operate power-plants than the land. Sea-water is a strongly corrosive to most metals, particularly in dry-and-wet splash zones. Small marine organisms are responsible for fouling, that is, attaching themselves to systems, thereby diminishing the systems' efficiency. And with depth the pressure increases, which tends to force sea-water into every seal and chamber. Marine architects and engineers are quite capable of dealing with corrosion, fouling and flooding, but the defences are costly.

These difficulties are not usually taken into account in economic analyses. Moreover, the basic cost data are not readily comparable due to changes in assumptions on construction, fuel cost, inflation, life-cycle length, methodology and immature technology. The projections are of course as uncertain as these assumptions. The cost estimates in Table 2 may be conservative, some may be in error by a factor of two or three, but they are the best estimates currently available.

TABLE 2. Economic projections

| Source | | Capital cost $/kW | Home-delivered cost[1] $/kWh |
|---|---|---|---|
| Existing utility plant | | | |
| Oil fired | 1,000 MW | 500 | 0.04 |
| Nuclear | 1,100 MW | 1 000 | 0.05 |
| Current | | | |
| CORIOLIS | 80 MW | 1 300 | 0.07 |
| Temperature | | | |
| OTEC | 250 MW | 2 400[2] | 0.07 |
| Wind | | | |
| MOD 1 | 2 MW | 1 600 | 0.09 |
| Offshore system | | 3 200[3] | 0.12[3] |
| Tide | | | |
| Rance | 240 MW | 1 000[4] | 0.05 |
| Maine | 500 MW | 3 500 | 0.09 |
| Salinity | | | |
| 100 MW range | | 4 000[3] | 0.10[3] |
| Wave | | | |
| 100 MW array | | 13 000[3] | 0.15[3] |

1. Generally $0.02 per kWh over at-plants costs.
2. Includes plant-factor limitation.
3. Denotes very uncertain estimates or averages.
4. 1968 cost, adjusted for inflation.

One calculates capital costs by adding up all expenses for materials, manufacture, assembly and installation. Dividing by the (rated) capacity of the generator then yields the plant cost per kilowatt. Systems costing up to $2,000 per kW are at present considered economic or near-economic.

A second calculation yields the costs per kilowatt-hour at which the power is sold (actually the sale is of energy, which is the product of power rate and time period) using an equation developed by P. B. S. Lissaman.

# The environment and the law

The use of ocean energy, like that of other renewable energies, is largely free of environmental impact. Where barrages are necessary, as with tidal and salinity plants, the restriction of migratory species, sediment transport and the flushing of pollutants from estuaries can be significant and damaging. The huge rotors of current and wind turbines could be a hazard to large seabirds and ocean mammals. Serious, widespread impacts can, however, be postulated only for levels of energy harvests that are significant fractions of the natural energy fluxes. This will not likely be the case since so-called downstream effects would discourage further deployment beyond certain levels before regional climates, for instance, could be altered.

If ocean power is used on-site in the manufacture of energy-intensive products by smelting, refining or synthesizing of materials, there will be the hazards normally associated with these processes: leaks, spills and waste disposal. These impacts may easily exceed the impact of the energy harvest itself.

The energy harvest will also introduce novel constraints into the current law of the sea. Some of the platforms may in time be moored in international waters for years on end, a practice without precedent. Although the nations of the world are seemingly unable to arrive at an acceptable international structure for the rights to ocean resources in the unratified Law of the Sea conferences, once the technology and economics of ocean energy harvest are worked out in territorial waters and 'economic zones', legal sanctions for use of open international waters should follow quickly.

# Outlook

The reader may ask, when, in which form and at what cost ocean energy will benefit him. If we trust the estimates in Table 2, then wave and salinity-gradient power may be the long shots, though wave work is avidly pursued in the United Kingdom, Japan and Sweden, and salinity work in Israel and the United States. The recently announced CORIOLIS turbine for the Gulf Stream current has surprisingly good prospects, but the low estimates may stem from early optimism about lack of difficulties. The only large existing ocean power-plant, the Rance tidal scheme in France, looks like an economic success, although that was not apparent during its construction; it was built from fuel-shortage considerations and not for cost advantage. At present, American tide projects do not appear so economically inviting. OTEC has middling prospects; it is the favourite in the United States and a 50-kW barge plant has begun tests off Hawaii. Onshore wind systems have low construction costs but medium operating costs and a number of countries are building large wind turbines; offshore, both construction and operating costs are likely to increase.

Within these systematic limitations, the reader must of course choose according to resource availability. Thus, CORIOLIS is mostly restricted to fast currents, OTEC to the tropics and near-shore deep water, wind turbines to the trades and westerlies, tide plants to converging shorelines, salinity plants to river mouths or salt deposits, and wave devices to generally stormy coasts.

The reader's energy recipe should take note of further ingredients. In the technologies dependent on exhaustible fuels, high efficiencies are of paramount importance. Technologies built upon renewable energy, however, could afford low efficiencies if they result in lower initial costs for alternative systems. Both efficiencies and costs are likely to improve with practice. As power production from renewable sources generally has little effect on the environment, a higher production density would be acceptable. Currently, as we are beginning to bring alternative sources on line in small amounts, the choice is wide and a source mixture would result from economic and geographic considerations. Continued growth in demand, however, will narrow the choice to the few technologies that can be deployed on a large scale. Concern has been expressed, moreover, that the critical factor in future energy supplies is having enough capital to develop them (Bockris, 1977). A goal of 10 kW per capita, about the present consumption in the United States, for everyone fifty years hence would require the investment of $4 trillion per annum, twice the annual gross domestic product of the United States. It is not likely to be achieved.

Emphasis on energy-and-power conservation thus appears to be an unavoidable part of bridging the gap between demand and supply. There are, of course, many strategies for conserving energy. Two that are particularly suitable for renewable sources are end-use matching and load management; that is, putting energy to its best or most direct use according to form, as mentioned earlier, and adjusting energy use to its fluctuating availability. A simple example of load management would be that of using electricity only when the windmill turns; a sophisticated one, that of cycling (running) of refrigerators activated by signals from the power-plant when the power

demand is low at night. Storing energy to even out supply is normally expensive and rarely cost-effective.

Ocean energy alternatives fit in well with the constraints and opportunities just discussed. As small-to-medium-scale systems they tend to be more distributed, closer to local demand and when linked together could average out fluctuations. This implies that they would be owned and operated locally, perhaps on a village or small-city level. And their manufacture, installation and operation would certainly create more jobs and work-places than is the case with today's large-scale power plants.

# References

BOCKRIS, J. O'M. 1978. Economic, Political and Psychosocial Barriers to a Change-over to Renewable Energy Resources. In: T. Nejat Veziroglu (ed.). *Alternative Energy Sources—An International Compendium*. Vol. 10, pp. 4903–17. Washington/London, Hemisphere Publ. Corp.

GUSTAVSON, M. R. 1979. Limits to Wind-power Utilization. *Science*, Vol. 204, No. 4388, 6 April, pp. 13–17.

ISAACS, J. D.; SEYMOUR, R. J. 1973. The Ocean as a Power Resource. *International Journal of Environmental Studies*, Vol. 4, pp. 201–5.

WICK, G. L.; SCHMITT, W. R. 1977. Prospects for Renewable Energy from the Sea. *Marine Technology Society Journal*, Vol. 11, No. 5/6, pp. 16–21.

# Biomass: solar energy as fuel

9

David O. Hall

*The biological conversion of solar energy via photosynthesis produces each year an amount of stored energy, in the form of biomass, about ten times the world's annual consumption. The amount of proven fossil fuel reserves underground equals the present standing biomass (mostly trees) on the earth's surface, while fossil fuel resources may be ten times this amount. The massive capture of solar energy and its conversion into a stored product occurs worldwide with an efficiency of only 0.1 per cent. The process occurs because of the adaptability of plants. It can be used over most parts of the earth.*

## Introduction

It is not widely appreciated that one-sixth of the world's annual fuel supplies are wood fuel and that about half of all the trees cut down are used for cooking and heating purposes. In the non-OPEC developing countries (which contain over 40 per cent of the world's population) non-commercial fuel often comprises up to 90 per cent of their total energy use. This non-commercial fuel includes wood, dung and agricultural waste and, because of its nature, is seldom thoroughly considered. That is to say total wood fuel consumption is probably three times that usually shown in statistics, and about half of the world's population relies

mainly on wood for cooking (four-fifths of total household energy use) and heating; statistics of non-commercial energy supply can be off by factors of ten or even a hundred times. (See Table 1.)

I would like to present some evidence that fuels produced by solar energy conversion are a very important source of energy now and will continue to be so for the foreseeable future—probably even to an increasing extent. The 'second energy crisis' has arisen as a result of diminishing supplies of non-commercial or traditional fuels such as wood, dung and straw. This has led to increasing deforestation, damage to ecosystems, and rural property. The prospects of fuels derived from biomass programmes are being considered in many industrialized and developing countries of the world. It seems to me, and others, that the biggest difficulty with deriving fuels from biomass programmes is that its simplicity belies its

Originally published as 'The Status of Solar Energy as Fuel', *Impact of Science on Society*, Vol. 29, No. 4, 1979, pp. 307–17.

TABLE 1. Fossil fuel resources, biomass production and balances of carbon dioxide, expressed in tonnes of coal equivalent (t) or joules (J)

| | Tonnes of coal equivalent | Joules |
|---|---|---|
| *Proven reserves* | | |
| Coal | $5 \times 10^{11}$ | |
| Oil | $2 \times 10^{11}$ | |
| Gas | $1 \times 10^{11}$ | |
| | $8 \times 10^{11}$ t | $= 25 \times 10^{21}$ |
| *Estimated resources* | | |
| Coal | $85 \times 10^{11}$ | |
| Oil | $5 \times 10^{11}$ | |
| Gas | $3 \times 10^{11}$ | |
| Unconventional gas and oil | $20 \times 10^{11}$ | |
| | $113 \times 10^{11}$ t | $= 300 \times 10^{21}$ |
| *Fossil fuels used to 1976* | $2 \times 10^{11}$ t carbon | $= 6 \times 10^{21}$ |
| *World's annual energy use* | | $3 \times 10^{20}$ ($5 \times 10^9$ t carbon from fossil fuels) |
| *Annual photosynthesis* | | |
| Net primary production | $8 \times 10^{10}$ t carbon $= 2 \times 10^{11}$ t organic matter | $= 3 \times 10^{21}$ |
| Cultivated land only | $0.4 \times 10^{10}$ t carbon | |
| *Stored in biomass* | | |
| Total (90 per cent in trees) | $8 \times 10^{11}$ t carbon | $= 20 \times 10^{21}$ |
| Cultivated land only | $0.06 \times 10^{11}$ t carbon | |
| *Atmospheric $CO_2$* | $7 \times 10^{11}$ t carbon | |
| *$CO_2$ in ocean surface layers* | $6 \times 10^{11}$ t carbon | |
| *Soil organic matter* | $10–30 \times 10^{11}$ t carbon | |
| *Ocean organic matter* | $17 \times 10^{11}$ t carbon | |

These data, although imprecise, show that (a) the world's annual use of energy is only one-tenth of the annual photosynthetic energy storage; (b) stored biomass is equivalent to the proven fossil fuel reserves; (c) the amount of carbon stored in biomass is approximately the same as the atmospheric carbon ($CO_2$) and the carbon as $CO_2$ in the surface layers of the ocean. Destruction of forests and soil humus releases about 4 to $8 \times 10^9$ t carbon to the atmosphere each year—equivalent to that released by burning fossil fuels.

*Source:* Grenon, Woodwell, Stuiver, Boardman and Pimentel, in D. Klass (ed.), *Symposium on Energy from Biomass and Wastes*, Chicago, Institute of Gas Technology.

credibility—such an approach seems too simple a solution to such a complex problem!

Besides the biological conversion of solar energy, there are significant long-term prospects for photochemical-photobiological systems to produce hydrogen, fixed carbon compounds, and electricity. Research in practical aspects of these fields has recently been rejuvenated and is proceeding well. These systems have important advantages over whole-plant systems and, if their problems can be solved, may become important solar-energy-for-fuel processes in the future.

A timetable for developing new fuel sources could be as follows:

1980–90: fuel from trees, other crops, residues; making more efficient use of existing biofuels; demonstrations and training for these.
1990–2000: increased use of complete crops and residues and the exploitation of local energy plantations.
2000 and beyond: intensive energy farming; improvement of plant species used; and artificial photobiology and photochemistry.

## The significance of photosynthesis

To return to photosynthetic conversion by plants, there are not many people today who need reminding that our fossil carbon reserves—whether for fuel or chemicals—are the products of past photosynthesis. Photosynthesis is *the* key process in life and as developed by plants can be simply represented as:

$$H_2O + CO_2 \xrightarrow[\text{solar energy}]{\text{plants}} \text{organic materials} + O_2$$

In addition to C, H and O, plants also incorporate nitrogen and sulphur into organic material via light-dependent reactions; this point is often not appreciated. Thus, the basic processes of photosynthesis have determined life as we know it (dependent on organic materials and oxygen) and will continue to play the major role in the functioning of bioenergetic systems in the future.

In the past photosynthesis has given us coal, oil and gas, fuel wood, food, fibre and chemicals. The relative use of these fixed carbon sources has varied over the years and will undoubtedly do so in the future. However, with today's abundant petroleum, the products of present-day photosynthesis are evident to the developed world mainly as food. We should re-examine and, if possible, re-employ previous systems; but, given today's increased population and standard of living, we cannot revert to old technology. We must develop new means of using present-day photosynthetic systems more efficiently.

Each year plant photosynthesis fixes about $2 \times 10^{11}$ tonnes of carbon with an energy content of $3 \times 10^{21}$ joules; this is about ten times the world's annual energy use and 200 times our consumption of food energy. The efficiency on land may be about 0.2–0.3 per cent overall, whereas agriculture may be about 0.5 per cent efficient (Table 2). It should be realized that these efficiencies represent stored energy and not just the initial conversion efficiencies so often quoted in other energy systems.

All the atmospheric $CO_2$ is cycled through plants every 300 years, all the $O_2$ every 2,000 years, and all the $H_2O$ every 2 million years. The magnitude and role of photosynthesis are largely unrecognized, principally because we use such a small fraction of the

David O. Hall

TABLE 2. Average-to-good annual yields of dry-matter production

| Climatic region | Tonnes/ hectare year | $g/m^2$ day | Photosynthetic-efficiency (percentage of total radiation) |
|---|---|---|---|
| *Tropical* | | | |
| Napier grass | 88 | 24 | 1.6 |
| Sugar cane | 66 | 18 | 1.2 |
| Reed swamp | 59 | 16 | 1.1 |
| Annual crops | 30 | — | — |
| Perennial crops | 75–80 | — | — |
| Rain forest | 35–50 | — | — |
| *Temperate (Europe)* | | | |
| Perennial crops | 29 | 8 | 1.0 |
| Annual crops | 22 | 6 | 0.8 |
| Grassland | 22 | 6 | 0.8 |
| Evergreen forest | 22 | 6 | 0.8 |
| Deciduous forest | 15 | 4 | 0.6 |
| *Savannah* | 11 | 3 | — |
| *Desert* | 1 | 0.3 | 0.02 |

*Source:* L. St Pierre (ed.), *Future Sources of Organic Raw Materials*, New York/Oxford, Pergamon Press, 1979 (proceedings).

fixed carbon and because we do not realize the importance of recycling phenomena. Any interference in the latter by pollution could have serious consequences.

## Input *v.* output energy

The increasing content of $CO_2$ in the atmosphere (a 25 per cent increase over the last 125 years) has become a worrying problem, especially if the world continues to burn fossil fuels at an increasing rate. Fortunately much has been published on this over the last two years (see Fig. 1) so that we are in a better position now to understand some of the factors involved, although we really need much more information before confident predictions can be made on the effects on the world's climatic patterns. There is no doubt, however, that renewable biological systems for fuel production will not contribute to the atmosphere's $CO_2$ concentration, and biomass may help alleviate some of the problems by acting as a temporary 'carbon sink'.

Plants are very adaptable and exist in great diversity; they could thus continue indefinitely to supply us with renewable quantities of food,

92

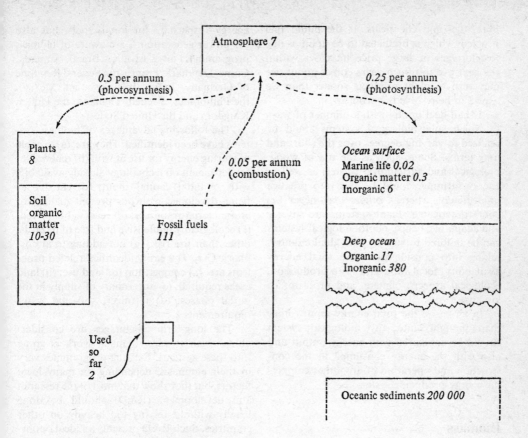

Atmosphere 7

0.5 per annum
(photosynthesis)

0.25 per annum
(photosynthesis)

Plants
8

Soil
organic
matter
10-30

0.05 per annum
(combustion)

Fossil fuels
111

Used
so
far
2

*Ocean surface*
Marine life 0.02
Organic matter 0.3
Inorganic 6

*Deep ocean*
Organic 17
Inorganic 380

Oceanic sediments 200 000

FIG. 1. The global carbon balance sheet showing
the main carbon deposits and annual exchange
rates, in units of $10^{11}$ tonnes. This simplified
diagram is adapted from Woodwell and Stuiver
and from Table 1. The total carbon reservoir
is about $430 \times 10^{11}$ tonnes, of which the
atmosphere represents 1.6 per cent, the biosphere
(plants, marine life and organic matter in soil
and ocean surface) 4.3 per cent
and the ocean 94.2 per cent.

fibre, fuel and chemicals. If the liquid fuel problem which is predicted to be upon us due to shortages or large price increases within the next five to fifteen years comes about, we may turn to plant products sooner than we expect to help solve the problem.

I shall deal briefly with a number of ways in which solar-biological systems could be realized to varying degrees over the short and long terms. Some, such as the use of wood, biological and agricultural wastes, as well as energy farming, could be put into practice immediately, whereas others may never become practicable. Plant systems are diverse and adaptable; hence, photobiological systems can be tailored to suit an individual country, taking into consideration the total energy availability, local food and fibre production, ecological aspects, climate, and problems of land use.

In all cases, the total energy input (other than sunlight) into any biological system should be compared with energy output and also with the energy consumed in the construction and operation of any other competing energy-producing system.

## Biomass

Solar energy is a very attractive source of energy for the future, but it does have disadvantages—what energy source does not! Solar energy is diffuse and intermittent on a daily and seasonal basis; thus, collection and storage costs can be high. However, plants are designed to capture diffuse radiation and store it for future use. Thus, there is much thought and money being given to ideas of using biomass (specially grown or as residues) as a source of energy—especially for liquid fuels, but also for power generation. I am aware of biomass programmes in Australia, Brazil, Canada, China, Denmark, France, the Federal Republic of Germany, India, Ireland, Israel, Mexico, the Philippines, Sweden, Thailand, the United Kingdom and the United States.

The following advantages of biomass systems have been identified. They are: (a) capable of storing energy for use at will; (b) renewable; (c) dependent on technology already available, with minimal capital input; (d) capable of being developed with our present manpower and material resources; (e) reasonably priced; (f) ecologically inoffensive and free of hazards, other than fire risk; (g) not adding to atmospheric $CO_2$. The easily identified related problems are: (a) competition for land use; (b) land areas required; (c) uncertainty of supply in the initial phases; (d) fertilizer, soil and water requirements.

The long-term advantages are considerable, which is why so much work is going into these systems. Existing programmes vary in their emphasis, depending on many local factors, but they show that most of the research and development (R&D) should be done locally without relying too heavily on other countries. Such R&D presents an ideal opportunity to develop and encourage local scientists, engineers and administrators in one field of energy supply. Even if biomass systems do not become significant suppliers of energy in a specific country in the future, the derivative benefits to agriculture, forestry, land use patterns and bioconversion technology are, I think, significant.

In Table 3, I have indicated the main biomass products.

TABLE 3. Simplified table relating processes, products and users to sun-generated biological conversion processes. Numerous cross-links exist; many important products and by-products are not listed. Agriculture is included in 'industry'

| Resource | Process | Products | Users |
|---|---|---|---|
| Dry biomass (wood and its residues) | Combustion | Heat, electricity | Industry, household |
| | Gasification | Gaseous fuels (methanol) Hydrogen, ammonia | Industry, transport, chemicals |
| | Pyrolysis | Oil, char, gas | Industry, transport |
| | Hydrolysis and distillation | Ethanol | Transport, chemicals |
| Wet biomass (sewage and aquatic biota) | Anaerobic digestion | Methane | Industry, household |
| Sugars (juices, cellulose) | Fermentation and distillation | Ethanol | Transport, chemicals |
| Water | Photochemistry-photobiology Catalysis | Hydrogen | Industry, chemicals, transport |

## The American programme

The United States has a very large R&D programme on biomass, with a budget of about $40 million for 1979. Details of the type of R&D being funded are available in numerous publications, but articles have appeared in *Science* that give invaluable details and references. A recent one by Burwell, 'Solar Biomass Energy: An Overview of U.S. Potential', gives a good general picture and presents some of his own ideas of the 'best' systems for the future. The American use of energy in 1975 was $71 \times 10^{18}$ joules, but the total standing forest inventory has an energy content three times this annual consumption. The total annual biomass growth of commercial forests is $9.3 \times 10^{18}$ joules, of which $6.6 \times 10^{18}$ joules is potentially collectible. Cropland agriculture produces energy total-

ling about $12 \times 10^{18}$ joules annually, of which about 40 per cent is represented by residues left on the land. Grain crops alone produce about $7.1 \times 10^{18}$ joules annually, of which $5.9 \times 10^{18}$ joules is the net collectible energy yield—$3.8 \times 10^{18}$ joules of this is in the form of residues. A detailed analysis of 'potentially usable biomass residues' shows that, of the residues currently collected, $2.1 \times 10^{18}$ joules could be obtained from urban solid wastes, $1.0 \times 10^{18}$ joules from animal feedlots, canneries, wood manufacture, and the like. Uncollected residues, such as cereal straw, cornstalks and logging waste, could contribute $5 \times 10^{18}$ joules per annum. Burwell considers that major opportunities for energy provision lie in the use of forest residues and improved management of productive forest land. He also makes the interesting point that 60 per cent of cropland in the United States is dedicated to the production of livestock; this excludes the contribution that 282 million hectares (38 per cent of the United States mainland) of pasture and rangeland make to support livestock. There are thus large areas of land which could be used in the future for the production of biomass, if society so wishes.

The concept of intensive silvicultural biomass farms (or energy farms) has been the subject of detailed analysis. Fast-growing deciduous trees which coppice (resprout from stumps) have been examined, as have numerous other trees, including those which fix nitrogen to ammonia. One recent major study came to some of the following conclusions:

Major energy products which could be economically derived from biomass at sometime in the future include electricity, ammonia, methanol, ethanol, and possibly medium-BTU fuel gas . . . the major opportunity for biomass in electric generation is in small plant retrofit or co-firing with coal . . . production of ammonia from wood biomass is estimated to be marginally competitive today . . . methanol production from wood could become competitive within the next decade.

An interesting fact which emerges from this detailed study on ten areas (six non-agricultural, two agricultural, one on swampland and one on forest) is that only 10 per cent of the land needs to be used for energy farming, so alleviating the necessity of acquiring 8- to 16-hectare blocks of land to fuel a power station—the limiting factor seems to be the distance required to transport the timber from the farm to the conversion facility.

Other studies indicate that a national total of 430 million dry tonnes are available each year. This figure is constituted by 280 million tonnes of crop residues, 120 million tonnes of residues from logging (at the mill and on the forest floor), and 30 million tonnes from animal manure. Excluded are residues from food processing, grazing animals, and hay and forage crops. Theoretically these 430 million tonnes of residue could supply 31 per cent of the electricity, 20 per cent of the natural gas, or 8 per cent of the oil requirements of the United States. This is very unlikely ever to happen, since at present 20 per cent of the residues are used and 58 per cent are returned to the soil—leaving only 22 per cent classified as excess. But these percentages will change as other uses (energy, chemicals) and prices develop.

In another study, Lipinsky outlines why he considers that fuels from biomass should be integrated with food and material production systems, i.e. adaptive systems should be encouraged which will 'modify themselves to meet evolving needs and constraints'. Such adaptive systems contrast with energy farms and the use of agricultural residues. Of course, the three systems can blend into each other, depending on circumstances. In this way biomass intermediates (besides food) are processed into fuels or materials, depending on relative price levels. Two examples cited are corn and sugar-cane production. Lipinsky concludes that

fuels from biomass may possibly supply 10 per cent or more of a curtailed [United States] energy consumption . . . competitive costs for fuels from biomass would depend primarily on the integration of fuel production with food and materials production . . . [and] knowing when and where to switch from emphasis on food and materials to emphasis on fuels is just as important as knowing how to produce the fuels.

# Other energy programmes

## Canada

Canadian studies on the large-scale production of methanol from biomass show that by 2025 between 4 and 42 per cent (depending on total energy use) of transport fuels could be provided by such methanol. 'Methanol represents a rather unique fuel combining the portability of liquid petroleum products and the clean, even-burning characteristics of natural gas,' one study says. It is shown that commercial production of methanol fuel would be feasible under certain conditions, and another analysis projects methanol costs of $0.35–$0.50 per gallon by the mid-1980s.

## Australia

Traditionally we think of energy plantations as forests, but increasingly we should consider alternatives, such as shrubs, weeds, agricultural crops, grasses and algae (freshwater and marine). In Australia five species have been selected, namely eucalyptus, cassava (*Manihot*), hibiscus, napier grass (*Pennisetum*), and sugar cane, as being potentially the most desirable high-yielding crops that can be harvested over the whole year. Recent calculations show that alcohol produced from starch-rich cassava is an economically viable system, but that if processing to destroy cell walls is required, the costs become too high. Methane and pyrolytic oil production from cereal straw and eucalyptus is calculated to be two to four times the equivalent cost of fuel in 1975 in Australia. If the prices of fossil fuels increase, the economics of photobiological processes will become more favourable, since fossil fuels

and electricity account for only 10–25 per cent of the cost of photobiological fuels.

## New Zealand

With its very efficient agriculture, low population density, and the fact that it uses a large proportion of its foreign exchange for petroleum purchase, it is natural that New Zealand is considering biomass as a source of fuels. Two studies are reported: alcohol from wood (pine trees) and biogas from agricultural crops (maize and kale). The biogas cost estimates are based on the use of 100,000-gallon digesters and processing 1,000 tonnes of dry matter/day.

## Philippines

In the Philippines a feasibility study has shown that a 9,100-hectare fuel wood plantation 'would supply the needs of a 75-MW steam power station, if it were not more than 50 km distant'. The investment requirements and cost of power produced look favourable and competitive with oil-fired power stations of similar capacity. Twenty-five such sites have been pinpointed, some of which could support power-plant capacities as high as 225 MW. The best species of fast growing trees seems to be the giant ipil-ipil (*Leucaena acidophila*), which fixes nitrogen to ammonia—a very desirable trait.

## Europe

In Europe a number of countries are conducting feasibility studies of the potential which biomass may have for supplying energy in the future. Trial plantings of alder, willow and poplar are being undertaken in addition to assessing energy yields from agricultural residues, urban wastes, techniques of conversion, waste land and forest potentials, and algal systems. Little has been published as yet, but a recent study (Project Alter) in France by Le Groupe de Bellevue proposes that in the long term France could produce liquid and solid fuels, comprising 11 per cent and 14 per cent of its total energy requirement respectively, from biomass sources. Land-use constraints will be a problem but, considering Europe as a whole, its past vegetative history with its diverse climates and land-use patterns, and its already burgeoning food surpluses, there may be far greater potential for biomass production than is commonly imagined.

A recent private study in Sweden, projecting energy supply and use in the year 2015, proposes that energy plantations could supply 351 terawatt hours (TWh), about two-thirds of the total consumed, from $2.9 \times 10^6$/ha (6 to 7 per cent of total land area) with an average production of 90 MWh/ha per year. Fast-growing willows and poplars seem to be the preferred species. A governmental energy commission believes that 55 TWh from biomass, of a total supply of 519 TWh, would be feasible in 1990.

## Brazil

By far the most ambitious biomass programme that has been planned is that in Brazil for the production of alcohol from sugar cane, sorghum, cassava and other crops. This national programme, called PNA or Proalcool, was established in November 1975. The alcohol produced will be used to blend with gasoline (petrol). Up to 20 per cent mixture (by volume)

requires no adjustment to engines. Over the last ten years the state of São Paulo, with more than 1.3 million cars, has varied the alcohol content of its petrol from 0.4 to 13.5 per cent—and 18 per cent in 1978—depending on the availability of alcohol and the price of molasses. Until August 1977, 141 new alcohol distilleries were authorized by Proalcool which would require an investment of about $900 million and would supply $3.2 \times 10^9$ litres of alcohol by 1980; this is about a fifth of the projected gasoline requirement. By 1985 total production of alcohol could reach 5 to $10 \times 10^9$ litres, with a total investment of about $3,150 million.

An economic analysis of the production of alcohol from sugar cane and cassava calculated selling prices of fuel, ex-distillery, of $333/m³ (this equals $0.33 per litre or $16.7/10^6$ BTU). These estimated prices are 81 per cent of the present retail price of gasoline on a volume basis, but are $43 to $73/m³ more costly than the present fixed-market price of alcohol of $290/m³; gasoline sells for $413/m³ or $13.8/10^6$ BTU. Thus the consumer is encouraged to use alcohol instead of gasoline, but the producer must receive an economic price in the future besides the government-guaranteed purchase of all biologically produced alcohol. Estimates of the energetics of alcohol production from sugar cane and cassava are favourable, since the energy output/input ratios have been calculated at between 6 and 9. These estimates may be somewhat high, but only large-scale agriculture and processing can finally determine the net energy ratio. What is clear is that Brazil is embarking on a programme of substitution of fuel imports using the natural advantages of land and climate which it has. This may be a very useful demonstration for other countries. The indirect benefits, such as saving foreign exchange, creating new employment, encouraging domestic technology and industry, and reducing pollution, are great.

### Sahel region

The problems of deforestation and desertification have highlighted the lack of fuel wood in most countries of this region. It is a scarcity which is reaching crisis proportions in large parts of South Asia, the Sahel, the Andean countries, Central America and the Caribbean.

The per capita requirement for cooking alone is about 0.5 m$^3$ of wood per year. Total fuel wood requirement is probably about 1 tonne, equivalent to about 400 kg of coal. A recent Netherlands study of the Sahel region points out two possible solutions: (a) decreasing fuel wood demand by using stoves which reduce consumption by 70 per cent; and (b) increasing the supply of fuel wood by establishing forest plantations and by converting wood into charcoal, since it is more efficient to use instead of wood—especially if fuel has to be transported over long distances.

### India

A study in Tamil Nadu, southern India, on the possibility of growing *Casuarina* (a nitrogen-fixing tree) as a source of fuel for a power-plant generating 100 MW electric (160 MW electric, installed capacity) has shown favourable economic and social benefits. In direct competition with coal, a payback period for such an energy plantation would be from fifteen to thirty years, depending on various factors. An equivalent of 110 hectares is required to generate 1 MW of electricity. About 11,000 labourers would be required.

### China

During the 1970s, biogas plants have been perfected and installed at a rapid rate. It is estimated that there were, in 1977, 4.7 million units in operation. The methane so produced is 'used for cooking, crop drying, power generation and various other purposes'. In Sichuan province alone, 17 million people use biogas for cooking and lighting; in some areas fully 80 per cent of all rural households are served by biogas.

## Photosynthesis in the future

One of the problems with photosynthesis is that it requires a whole plant (or alga) in order to function, and the problem with whole-plant photosynthesis is that its efficiency is usually low (less than 1 per cent). Many limiting factors of the environment and the plant itself interact to determine the overall efficiency. The most easily identified limiting factors are high light intensities, high and low temperatures, $CO_2$ concentration, water availability, supply of nutrients (especially nitrogen), availability of sinks for the products of photosynthesis, and so on. Knowing how these factors operate individually is obviously important, but what seems much more important is to try to understand how they interact in determining whole-plant yields. This is an immense task, but is worthwhile tackling, since plants produce a stored product at a seemingly low efficiency.

Thus, a task for photosynthesis of the far future is to try to select or manipulate plants which will give higher yield (of biomass, fuel, fibre, chemicals or food) with acceptable energy output/input ratios. We need to place much more effort on studies of whole-plant physiology and biochemistry and their interactions with the environment. This type of research is being increasingly funded by both industrial and governmental organizations who see a good future for plant-based systems. In the past, research in the plant sciences has been a poor relation in the scientific world—it has been taken for granted far too long. Now many questions, seemingly simple, are being asked to help solve problems of plant productivity in different environments. We have few answers, and it takes time to get them because of the lack of basic knowledge.

## Man-made photosynthesis and photochemistry

Since whole-plant photosynthesis operates with many limiting factors, would it be possible to construct artificial systems which mimic parts of the photosynthetic process and produce useful products at higher efficiencies of solar energy conversion? A 13 per cent maximal efficiency of solar energy conversion is considered a practical limit to produce a storable product. I think that this is definitely feasible from a technical point of view, but it will take time to discover if it could ever be economic. Note must also be taken of other chemical and physical systems, such as light-driven processes, that are being investigated and may come to fruition before biologically based systems do.

Plants are the centre of at least two unique reactions upon which all life depends and which have not yet been emulated in artificial systems, namely the splitting of water by visible light to produce oxygen and protons (hydrogen) and the fixation of $CO_2$ in organic compounds. Understanding how these two systems operate and mimicking the processes *in vitro* to produce $H_2$ or organic carbon are now the subjects of active research by photobiologists and chemists.

## A few final words

In conclusion, I reiterate that photosynthesis is the key process in the living world and will continue to be so for the continuation of life. The development of photobiological energy conversion systems has long-term implications from the energy, food, fibre, and chemical points of view. The applicability of these systems might be immediate in some tropical and subtropical areas and countries with large amounts of sunshine. Whatever systems are devised in the temperate zones could also be applicable to those countries with more sunshine (these are predominantly the developing countries). Thus the temperate countries could help themselves by becoming more self-sufficient and help other nations by not competing unnecessarily for their own fuel, food and raw materials. Lastly, we might find an alternative way of providing ourselves with food, fuel, fibre and chemicals in the next century. We should consider all our energy options and not invest everything in one or two energy systems as we have in the past.

# Hydro-power   10

United Nations Environment Programme

*Direct use of water-power (waterwheels) dates back to early civilizations; this peaked (as water-powered mills) in the nineteenth century. Hydro-electric production of energy then began to develop and today accounts for about 20 per cent of total electricity production. A review of world hydro-power resources by regions shows marked underdevelopment of this resource in the developing regions (only 4 per cent of potential in Africa, 22 per cent in Asia and 27 per cent in Latin America). Caution is advised in installing large-sized generating plants in developing countries where associated patterns of distribution and use can distort general societal development. Greater attention to small hydro-power stations may prove a wiser course for the present.*

Man's earliest extensive use of energy, other than muscle-power of man and animals and direct solar energy, was that derived from flowing water. The use of waterwheels of various types extends back to the early civilizations. The size and efficiency of waterwheels increased over the centuries, and in the nineteenth century, water-powered mills of various types ushered in the beginning of the industrial age. The peak of this early water-power development phase was reached about the middle of the nineteenth century because favourable mill sites, within the reach of the mechanical transmission of power, were limited. Furthermore, at that time, the more flexible steam-engines were improving in economy and de-

pendability. With the advent of electricity in the 1880s, and with alternating-current technology making transmission of electric energy more economical, the development of hydro-electric energy was well under way by the beginning of the twentieth century. Developments were rapid and by the 1930s projects such as the 1.3-million-kW powerhouse at Hoover Dam in the United States were completed. Large hydro-electric installations such as this increased the utilization of energy in the industrialized countries, and programmes to utilize the large hydro-electric potentials were pushed ahead.

The growth of electricity production from hydro-power has considerably increased since the 1950s. In 1950, the hydro-electricity production was $343 \times 10^9$ kWh; in 1978 it reached $1,557 \times 10^9$ kWh, i.e. an increase of 4.5 times.

Edited version of Chapter VI: 'Hydro-power', UNEP Report, 1980.

Table 1 gives the hydro-electricity production as compared to electricity produced from other sources. On the average, hydro-electricity constitutes about 21 per cent of the total world electricity production.

TABLE 1. World electricity production (in $10^9$ kWh)

| Year | Thermal | Hydro | Nuclear | Total | Percentage Hydro |
|---|---|---|---|---|---|
| 1972 | 4 226 | 1 294 | 144 | 5 664 | 22.8 |
| 1973 | 4 570 | 1 317 | 197 | 6 084 | 21.6 |
| 1974 | 4 567 | 1 446 | 255 | 6 268 | 23.1 |
| 1975 | 4 660 | 1 467 | 351 | 6 478 | 22.6 |
| 1976 | 5 063 | 1 471 | 410 | 6 944 | 21.2 |
| 1977 | 5 227 | 1 524 | 509 | 7 260 | 21.0 |
| 1978 | 5 527 | 1 557 | 530 | 7 614 | 20.5 |

*Source:* United Nations (1979).

Table 2 gives an estimate of the potential of the world's hydro resources together with the calculated percentage of developed resources in the different regions. Armstrong (1978) estimated the total potential from hydro resources of the world at $2.2 \times 10^6$ MW of installed and installable generating capacity at 50 per cent capacity factor. It should be noted that these estimates are conservative and do not take fully into account the great potential of resources for small hydro-power schemes.

The operating hydro-electric capacity at present is about 400,000 MW with an annual production of about $1,558 \times 10^9$ kWh, which is approximately 35 per cent of the total world potential available 95 per cent of the time. Although about 75 per cent of the hydro potential in Europe and North America has been exploited, only about 22 per cent in Asia, 27 per cent in Latin America and 4.3 per cent in Africa has been harnessed for energy production.

TABLE 2. World hydro-power resources

| | Potential available 95% of time ($10^3$ kW) | Potential output 95% of time ($10^6$ kWh/y) | Present installed capacity ($10^3$ kW) | Current annual production ($10^6$ kWh/y) | Percentage of developed potential (4)/(2)$\times$100 |
|---|---|---|---|---|---|
| | (1) | (2) | (3) | (4) | (5) |
| Africa | 145 218 | 1 161 741 | 11 437 | 49 663 | 4.3 |
| Asia | 139 288 | 1 114 305 | 59 773 | 245 096 | 22.0 |
| Europe (incl. USSR) | 102 961 | 827 676 | 177 797 | 620 676 | 75.0 |
| North America | 72 135 | 577 086 | 111 402 | 434 035 | 75.2 |
| Latin America | 81 221 | 649 763 | 38 582 | 176 845 | 27.2 |
| Oceania | 12 987 | 103 897 | 9 578 | 31 669 | 30.5 |
| TOTAL | 553 810 | 4 434 468 | 408 569 | 1 557 984 | 35.1 |

*Sources:* Columns (3) and (4) are 1978 figures according to United Nations (1979). Columns (1) and (2) are after UN Water Conference Secretariat (see Biswas, 1979).

Hydro-power is an important renewable source of energy, and constitutes an integral part of optimum overall water-resource utilization. It is a catalyst in socio-economic development, particularly in rural areas of developing countries (El-Hinnawi, 1977a, 1979); and its economic justification is improving because of its 'inflation-proof' characteristics, and its long life and low maintenance costs (Armstrong, 1978). It should be noted, however, that in order to maximize the benefits of hydro-power, it should be developed as a part of an overall water development plan. The development of large-scale water-power *per se* in some developing countries, for example, has created a number of socioeconomic problems. One consideration is that the output of electricity requires, for example, expensive transmission lines and a societal and economic structure ready to take advantage of this form of energy (Kristoferson, 1977). Low density of population, long distances between energy consumption centres, and low income levels are some factors that often hamper the growth of electricity consumption among the population at large. Because of the large investments necessary for many dams, in many cases only very large-scale industrial operations are regarded as economically viable. This may result in large, energy-intensive industries being constructed to match the output of large power stations. The influence of this type of industrial development on the general societal development of countries is sometimes questioned. Adequate planning and cost-benefit analysis is, therefore, an important prerequisite for the development of hydro-power installations.

Hydro-electric generating plants vary considerably in size, from as small as 3 kW (El-Hinnawi, 1977a) to as large as $12 \times 10^6$ kW (ICOLD, 1977). Small hydro-power stations (also called mini-hydro, village-size hydro-power installations, etc.) are generally those with installed generating capacity less than 1,000 kW (El-Hinnawi, 1979).[1] Such stations are normally built as part of water-management and rural-development strategies. In China, for example, more than 60,000 small hydro-power stations, with a total generating capacity of about 3,000 MW, were built in the last fifteen years (El-Hinnawi, 1977b) in rural areas. This has largely accelerated the development of such areas by providing power for better irrigation schemes, small-scale agricultural and rural industries, and domestic use. There is no technical reason why such small hydro-power schemes should not be built in rural areas in other developing countries using any suitable available water course (canals, river, reservoirs, falls, etc.).

Large hydro-electric power-plants can reach more than 12,000 MW (for example, the Itaipu power-plant on the Brazil–Paraguay border, which is under construction). According to ICOLD (1977), there are about 100 plants exceeding 1,000 MW in the world (29 became operational in the last decade; 36 are under construction or in early phases of development; the rest became operational before 1970). The construction of such large hydro-electric power-plants occurs on major rivers, and the reservoirs resulting from build-

1. Smil (1976) defines small hydro stations as those with capacities less than 500 kW, whereas in China, small hydro-power schemes are those having installed generating capacities less than 12,000 kW (El-Hinnawi, 1977b). In the United States small hydro is a unit generating less than 15,000 kW (Wehlage, 1979).

ing the dams vary in area from one place to another. There are about forty man-made reservoirs that range from 1,000 km² to 8,730 km² (Fels and Keller, 1973). The volumes of the reservoirs vary from one site to another and can reach about $200 \times 10^9$ m³ (Mermel, 1976). Table 3 gives some examples of large hydro-electric power-plants and the area of the associated reservoirs.

TABLE 3. Examples of large hydro-electric power-plants and of large reservoirs

| Dam | Present installed capacity (MW) | Ultimate installed capacity (MW) | Reservoir area (km²) |
| --- | --- | --- | --- |
| Krasnoyarsk (USSR) | 6 096 | 6 096 | 2 130 |
| Churchill (Canada) | 5 225 | 5 225 | 6 200 |
| Bratsk (USSR) | 4 100 | 4 600 | 5 426 |
| Volgograd (USSR) | 2 560 | 2 560 | 3 160 |
| Volga Lenin (USSR) | 2 300 | 2 300 | 6 500 |
| Aswan High Dam (Egypt) | 2 100 | 2 100 | 5 250 |
| Saratov (USSR) | 1 360 | 1 360 | 1 950 |
| Kariba (Zimbabwe/ Zambia) | 1 266 | 1 866 | 5 180 |
| Furnas (Brazil) | 1 216 | 1 216 | 1 606 |
| Kainji (Nigeria) | 320 | 1 000 | 1 243 |

Sources: ICOLD (1977), Fels and Keller (1973).

Different types of dams ranging from earth to concrete with different designs have been built. In some cases, particularly in small hydro-power schemes, it is not necessary to build dams. The development of bulb-type, self-contained turbine-generators can utilize the normal flow of rivers in some cases. It should be noted that there are about 24,000 dams in the world and that dams are not necessarily built for harnessing hydro-power; for example, in the United States dams are being built at the rate of 125 per year, about 5 per cent of which are associated with hydro-power, the rest with water supply, irrigation, etc. (Mermel, 1976).

Hydro-electricity generation has a number of environmental impacts. No dam can be built and no lake can be created without environmental costs and benefits of some kind. A dam becomes a dominant factor in the hydrological regime, and sets in motion a series of impacts on physical, biological and socio-cultural systems. The many consequences on the environment of the dam and the lake behind it appear to be factors in common regardless of the dam's geographical location. The environmental side-effects of dam construction are generally divided into two categories: (a) the local effects and the reactions within the area of the man-made lake; (b) the downstream effects resulting from a change in the hydraulic regime. Both categories have their physical, biological and socio-economic elements.

Conventional hydro-electric developments use dams and waterways to harness the energy of falling water in streams to produce electric power. Pumped storage developments utilize the same principle for the generating phase, but all or part of the water is made available

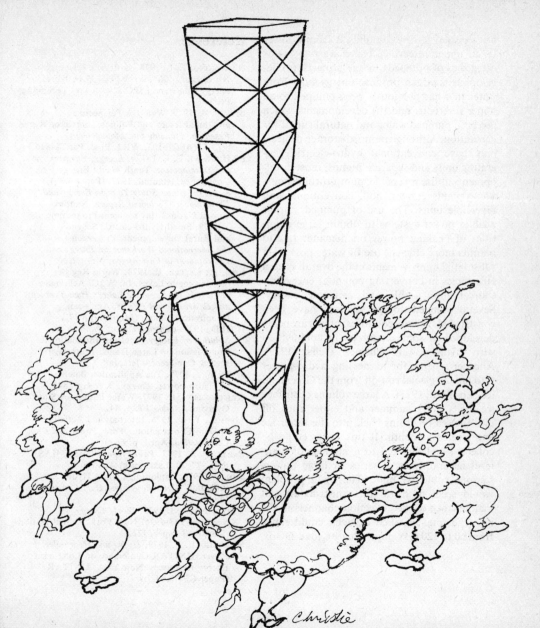

Christie

for repeated use by pumping it from a lower to an upper reservoir. There are two major categories of pumped-storage projects: (a) developments which produce energy only from water that has previously been pumped to an upper reservoir, and (b) developments which use both pumped water and natural runoff for generation. Although pumped-storage projects may have conventional hydro-electric generating units and separate pumps, most developments utilize reversible, pump-turbine units. Some plants contain both conventional and reversible units. The use of pumped storage enables power systems to obtain large quantities of peaking power on demand. It also permits more efficient use of water power via substantial improvements in the overall system efficiencies in converting potential energy resources into usable and available energy. Several pumped-storage schemes have been constructed in many countries; others are being developed in some countries in conjunction with hydro-power schemes (Jeffs, 1979). Another scheme for harnessing hydro-power is to tap the glacial run-off from the Greenland ice-cap (Partl, 1977). A large volume of surface ice melts every summer and either runs off into the sea or drains back into the lower ice and freezes at depth. If this water could be collected and directed to a number of high-level natural lakes, a series of large power-stations could be built. The main problem would seem to be in the construction and maintenance of the collection channels in the ice. The total installed capacity could range from 60 to 120 GW depending on load factor.

# References

ARMSTRONG, E. L. 1978. Hydraulic Resources. *World Energy Resources 1985–2020. World Energy Conference.* IPC Science and Technology Press.

BISWAS, A. 1979. Water: A Perspective on Global Issues and Politics. *Journal of Water Research, Planning and Management,* p. 205. (ASCE 105, WR2, Proc. Pap. 14815.)

EL-HINNAWI, E. E. 1977a. *Energy, Environment and Development.* Tenth World Energy Conference, Istanbul, 1977. (Paper 4.8–1.)

——. 1977b. *China Study Tour on Energy and Environment. Technical Report.* Nairobi, United Nations Environment Programme.

——. 1979. Small Hydro-electric Schemes and Rural Development. *Proceedings of the International Workshop on Energy and Environment in East Africa, Nairobi.*

FELS, E.; KELLER, R. 1973. World Register on Man-made Lakes. In: W. C. Ackerman et al. (eds.), *Man-made Lakes; Their Problems and Environmental Effects.* American Geophysical Union. (Monograph 17.)

ICOLD, 1977. *Contributions of Dams to the Solution of Energy Problems.* International Commission on Large Dams, Tenth World Energy Conference, Istanbul.

JEFFS, E. J. 1979. The Application Potential of Hydro-power. *Energy,* Vol. 4, p. 841.

KRISTOFERSON, L. 1977. Water Power—A Short Overview. *Ambio VI,* p. 44.

MERMEL, T. W. 1976. International Activity in Dam Construction. *Water Power and Dam Construction,* April, p. 66.

PARTL, F. R. 1977. *Power from Glaciers.* IIASA, R.R.–77–20. Laxenburg, Austria.

SMIL, V. 1976. Exploiting China's Hydro Potential. *Water Power and Dam Construction,* March, p. 19.

UNITED NATIONS. 1979. *World Energy Supplies 1973–1978.* New York, United Nations. (Statistical Paper, J22.)

WEHLAGE, E. F. 1979. *Hydro and Hydro-electric Power.* UNITAR Conference on Long-term Energy Resources. New York, UNITAR. (Paper CF7/XVI/2.)

# Microbes for energy

## 11

Edgar J. DaSilva

*Current patterns and practices of industry and agriculture in most developed countries
generate enormous amounts of organic wastes: agricultural residues, domestic garbage,
industrial effluents and the like. These countries are also in need of energy. The developing
countries, in their quest for technological parity, have been trekking the same routes towards
industrialization. Corollary factors are those of (expensive and dwindling) fossil energy
resources, unemployment and economic stagnation. In the search for an environmentally sound
and less energy-intensive approach, a mix of several microbiological processes appears to
offer an attractive solution. This offers the ancillary benefits of increased employment
opportunities and a rejuvenated economy.*

The vast potential of microbial technology in catalysing economic progress, especially in the developing countries, in terms of substitute sources of fuel, food, fertilizer and feed supplements has already been documented (Da Silva et al., 1978; Bull et al., 1979).

In the developing countries, traditional farming is supported by photosynthesis, a natural solar-energy conversion mechanism sustaining life from the primitive to the most advanced. Additional energy sources include the use of organic waste and the deployment of human and animal power for raising crops. Agriculture in the industrialized countries is highly protected as well as highly productive. Consequently, costs are high; this results from a mesh of factors—additional agricultural inputs through the use of fertilizers, pesticides, improved irrigation and drainage methods, and fossil-fuel energy in improved machinery and farm management practices (Table 1).

The role of micro-organisms as tools for rural processing of organic residues has recently been discussed (Porter, 1978a), and several examples have been provided showing the deployment of micro-organisms for the production of the five Fs: fuel, fibre, fertilizer, feed and food (Fig. 1). La Riviere (1978), in focusing

Originally published as 'Micro-organisms
as Tools for Biomass Conversion and Energy
Generation', *Impact of Science on Society*, Vol. 29,
No. 4, 1979, pp. 361–74. It was dedicated to
the late Professor J. R. Porter (University of
Iowa, Iowa City), former chairman
UNEP/Unesco/ICRO Panel on Microbiology,
in recognition of his work in the field of
microbiology and its allied disciplines.

TABLE 1. Fossil energy inputs into agriculture

| Energy-dependent inputs (1972/73) | Total world consumption (in million tonnes) | Percentage consumption (developed countries) | Percentage consumption (developing countries) | 1985 projections (global) (in million tonnes) | 1985 percentage projection (developing countries) | 1985 percentage projection (developed countries) | Fossil energy requirements per kilogram of energy-dependent input |
|---|---|---|---|---|---|---|---|
| Nitrogen fertilizer | 36.00 | 83.3 | 16.7 | 85.0 | 25 | 75 | 2.00 |
| Phosphate fertilizer | 22.00 | 86.0 | 14.0 | 42.5 | 20 | 80 | 0.25 |
| Potash fertilizer | 18.75 | 90.0 | 10.0 | 38.0 | 17 | 83 | 0.18 |
| Insecticides and herbicides[1] | 1.60 | 94.0 | 6.0 | 2.3 | 23 | 77 | 2.40 |

1. 1970 figures.

*Source:* 'Energy for Agriculture in Developing Countries', *Science and Public Policy*, April 1977, pp. 168–78.

on environmental goals for microbial conversion in rural communities, has emphasized that the incentives for energy recovery from wastes are strongest in the rural areas of the developing countries.

Agricultural and forest residues have been increasingly investigated as a possible source of energy. The advantages of this potential energy are domestic production, renewable supply, and clean conversion. In addition, the bioconversion of organic and agricultural residues is growingly attractive because it has been proven by scientific and technological experience. Non-polluting in its technology and feeding upon renewable resources, bioconversion of organically biodegradable resources is triply appealing: it has roles in energy production, counteracting deterioration of the environment, and waste management.

In what La Riviere terms 'from sense to non-sense; and back to sense again', the case is made for evaluating situations where prices of energy and fertilizer are mounting—thus necessitating a reorientation of the available processes involving micro-organisms to bring within easy reach of the household the recovery of energy and allied essential attributes. Microbes are not only efficient at capturing solar energy; they are equally important in expanding agriculture from land into water through aquaculture and even ocean farming. Thus integration has a twofold purpose: the integrated use of all types of living organisms also makes it possible to integrate food, fodder, fuel, fibre and chemical production into a new type of highly economic, multi-purpose agriculture. In producing more food for the planet's increasing population and in sustaining technological growth and progress, it is clearly evident that deep inroads will continue to be

The basics of bio-energy

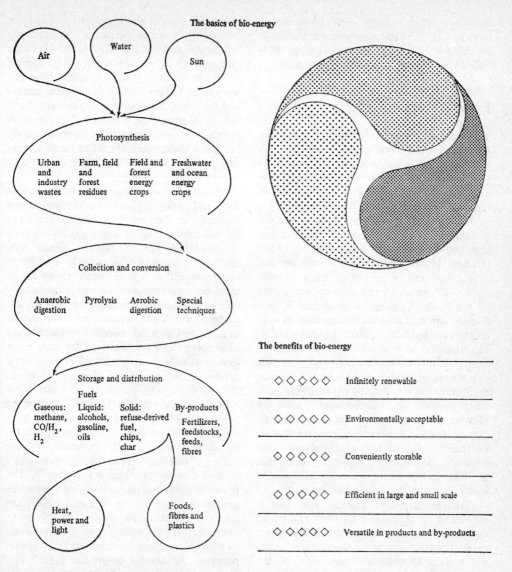

The benefits of bio-energy

| | |
|---|---|
| ◇ ◇ ◇ ◇ ◇ | Infinitely renewable |
| ◇ ◇ ◇ ◇ ◇ | Environmentally acceptable |
| ◇ ◇ ◇ ◇ ◇ | Conveniently storable |
| ◇ ◇ ◇ ◇ ◇ | Efficient in large and small scale |
| ◇ ◇ ◇ ◇ ◇ | Versatile in products and by-products |

FIG. 1. Basics and benefits of bio-energy.

made into nature's energy bank. So to change daily life-styles and programmed industrial growth patterns in midstream (though possible) is not an easy and attractive solution.

## From biomass to energy

The four main sources of energy are: (a) sun, water and wind (indefinite and freely available); (b) renewable resources (non-fossil organic materials—e.g. straw, human and animal power); (c) finitely available fossil fuels (coal, oil and gas); and (d) nuclear energy.

Increased attention is being focused on the possibilities of using a regenerable resource, high quality biomass as a fuel (Porter, 1978b; Gruber, 1978). The United States extracts 1 to 2 per cent of its energy from biomass ($3.02 \times 10^{15}$ joules of the national energy use of $174.37 \times 10^{15}$ joules), and optimistic predictions are that this would increase to 6 to 25 per cent by 2020 to 2025. Lipinsky (1978) has analysed several adaptive schemes that deploy biomass—sugar cane, corn, guayule as a source of food, materials and chemicals, in addition to the development of safe nonpolluting fuels. Biomass-to-energy systems are of three types: those involving direct combustion of biomass materials to generate heat or electricity, those involving animals or other livestock and those that yield molecules that are richer in high energy elements (carbon and hydrogen) than the original biomass.

Transformation of biomass into fuel is influenced by geographic location and by an optimal process of energy conversion. Biomass production happens to be the greatest in the wet equatorial regions rather than in the northern temperate lands where fuel consumption is the highest and frequently wasteful. In developing countries, the potential for biomass production and conversion is much higher than current levels of fuel consumption. The limiting factors in harnessing this potential, however, seem to be the acquisition of sufficient capital and the timely development of adaptable technology. A great portion of the energy bound in biomass is lost during conversion to utility fuels; but these losses appear to be no greater than those involved in the conversion of coal into synthetic oil and gas.

Wastes from non-energy sources, such as food and paper production and crops grown explicitly for their energy value, are known to be valuable sources of organic fuels. Defined as 'resources out of place', organic biodegradable wastes contain energy that is recoverable either by physical, chemical or microbiological means. Energy is physically recovered through incineration of sewage sludges, municipal refuse and solid wastes of animals. Chemical processes involve the use of pyrolysis and gasification. The most common method—a microbiological non-waste producing technology—is biogas production, the problems and prospects of which have recently been reviewed (National Academy of Sciences, 1977; DaSilva, 1978) and the technology of which ranges from simple, small-farmer lagoons through the medium technology of bag-digesters to the high technology of high-rate biogas digesters (Table 2). The principles involved in methane or biogas digesters are very simple. Organic wastes are decomposed, through the action of methanogenic bacteria within a sealed container, to provide methane for cooking and lighting purposes. A valuable by-product is fertilizer. Several developing countries use biogas plants

TABLE 2. Comparison of bioconversion plants

| Size and type of plant | Location | Daily estimated methane gas output (in ft$^3$ CH$^4$/ft$^3$ of digester)[1] | Construction materials and design type | Estimated procurement and construction time | Builder |
|---|---|---|---|---|---|
| Plug-flow | China | 0.4–0.5 | Concrete without plumbing | 2 weeks to 3 months average: 1 month | Local peasants |
| Batch-flow | India | 0.75 | Steel gas dome with plumbing | 1 to 4 months average: 2 months | Local peasants |
| 75 cow, plug-flow | Ithaca, N.Y. | 1.5 | | 9 months | Dr Bill Jewell, Cornell University for Department of Energy |
| 75 cow, complete mix | Ithaca, N.Y. | 1.5 | | 9 months | Dr Bill Jewell, Cornell University for Department of Energy |
| 300 cow, complete mix | Monroe, Washington | 1 minimum 4 maximum | A. O. Smith Slurrystore tanks, complex expensive sewage treatment machinery | Actual 6 months; 3 months w/standardized drawings, union construction | Ken Smith, Ecotope Group State of Washington and Department of Energy |
| 400 cow, plug-flow | North Glen, Colorado | 2.5 | | 3 months | Energy Harvest, Incorporated |
| 1,000 household, controlled landfill | Proposed | 2 | Polyethylene plastic, pipe and scrubbers | 12 months | Proposed (K. Smith, Dynatech R&D) |
| 10,000 home, Ref CoM 'garbage-to-gas' | Pompano Beach, Florida | 3 | | 16 months | Waste Mgmt, Incorporated for Department of Energy |

1. 1 ft$^3$ equals 0.02832 m$^3$ and 1 in$^3$ equals 16.39 cm$^3$.
Source: L. de Moll and G. Coe (eds.), Stepping Stones—Appropriate Technology and Beyond, New York, Schocken, 1978.

FIG. 2. Biogas map showing R&D activities and implementation in developed and developing countries.

and have several integrated schemes functioning (Fig. 2).

## Some specific conversions

In Africa, there is increasing interest in the utilization of biogas plants. In Kenya, biogas plants have been used since 1954. From 1955 on Tunnel Estate, Fort Ternan, biogas has been the only fuel used for cooking. Further-more, coffee has been grown entirely on the fertilizer residues of the methane plant. Active experimentation in harnessing the multiple utility benefits of biogas technology are currently being carried out in the United Republic of Cameroon, Ethiopia, Rwanda, Senegal, United Republic of Tanzania, Upper Volta, Zaire and Zambia. Approximately 38 per cent of the world's 9,500 million head of livestock are in Africa, Asia, Latin America and the Near East. In addition to the enormous amount of manure thus made available, it has been estimated (in 1974) that almost 95 per cent of the world's total residues from bananas,

TABLE 3. Fuels from biomass

| Organization | Title | Projected contribution |
|---|---|---|
| Agricultural Research Service | Anaerobic fermentation of livestock manures and crop residues | Recovers high-protein biomass methane and plant nutrients from livestock manures and crop residues through anaerobic processes |
| Cornell University | Anaerobic fermentation of agricultural residues; potential for improvement and implementation | Optimizes fermenter designs for use in large-scale pilot plants |
| Dynatech R&D Co. | Engineering evaluation of programme to recover fuel from gas waste | Evaluates various methods of methane production, and monitors technical anaerobic digestion projects |
| Ecotope Group | Monitor, test and evaluate an operational digester for 350 head of cattle | Applies knowledge of digester applications to full-scale dairy farm operations |
| Hamilton Standard | Feedlot energy reclamation demonstration | Produces fuel gas to achieve self-sufficiency of feedlot energy |
| Hamilton Standard | Pipeline fuel gas from environmental cattle feedlot | Produces fuel gas and other products from animal residues; promotes commercialization of process |
| Illinois, University of | Biological conversion of biomass to methane | Converts organic refuse to methane in an efficient, cost-effective manner |
| Stanford University | Heat treatment of organics for increasing anaerobic biodegradability | Increases biodegradability and methane production from waste organic materials by heat treatment under pressure |

*Source:* R. Ward, 'U.S. Digestion Research and Development and Demonstration Programmes and Application to Asia and the Pacific', Expert Group Meeting on Bio-gas Development, ESCAP, Bangkok, 20–26 June 1978. (Report.)

citrus fruits, cassava and coffee are in Africa, Latin America and Asia.

Also found in these regions is 72 per cent of the world's total residues from sugar cane. In the United States alone, estimated quantities of methane gas which can be produced from nearly 881 million tonnes of dry organic matter, available on an annual basis from known waste sources, amount to $408 \times 10^9$ m³. In 1976, the Federal Power Commission authorized Calorific Recovery Anaerobic Process Inc. of Oklahoma to provide the Natural Gas Pipeline Company annually with 23 million m³ of methane derived from feedlot wastes.

The United States Department of Energy

(DOE) has sponsored a number of research development and demonstration projects (Table 3) on the utilization of organic wastes and crops especially grown for conversion to clean fuels. The underlying principles of all DOE studies are the ready availability of large quantities of organic residues, the displacement of petroleum consumer products by methane gas and its ancillary benefits. The last include production of feed supplements, pollution control and recycling of nutrients that result from the process technology. Most biomass-derived fuels come from regenerable crop energy resources, constituting appropriate technological alternatives in the drive to reduce dependency on fossil fuels. But the disadvantages of the use of biomass in its various forms and its conversion to synthetic fuels are low heating value, high water content, and high costs of harvesting and transport. Nevertheless, biomass popularity is increasing and indicators of such use include the following:

Fiat has developed a high-efficiency methane-fuelled electrical generator based on a standard automobile engine, for use as a source of household energy, 'Totem' (total energy module).

Countries such as Australia, Canada, Sweden and the United States, in particular, envisage meeting a rising proportion of their energy needs through bio-energy systems.

A draft outline for a future study on 'Waste Recycling Technology for Development' is to be considered by the United Nations Advisory Committee on the Application of Science and Technology to Development (ACAST) for further action (United Nations, 1978). Proposals for case studies that were recommended dealt with waste recycling in Kuwait, product and energy recovery from the residues of ethyl alcohol from fermentation and distillation, recycling agricultural wastes in Kenya and other African countries, new trends in the creation of non-waste technology in the Soviet Union, and solid-waste utilization in India.

## Alcohols as fuel substitutes

The energy policy in Brazil is geared to intensifying the use of locally abundant primary sources, increasing available reserves, rationalizing consumption and production rates, and reducing dependence on petroleum as the main source of primary energy. A national alcohol programme has been established, with the view to substituting ethanol for petrol (gasoline) and diesel oil and increasing the production capacity of ethanol in balance with a reduction in petroleum imports. The utilization of alcohol in Brazil can be traced back to 1931, when the then government decreed an ethanol addition of 5 per cent to imported gasoline as a means of saving a sugar industry bedevilled by a decline in sugar prices and increasingly inaccessible export markets.

Elsewhere too, ethanol production by fermentation was deployed for fuel production, particularly in Europe after the First World War. In the United States, Agrol, consisting of 90 per cent gasoline and 10 per cent anhydrous alcohol, was manufactured in 1935. Today, Gasohol is being widely promoted, but the economics of manufacturing and processing the product still remains to be established conclusively in terms of its comparison to the economics of normal petrol.

At the University of Queensland scientists are engaged in research aimed to improve the

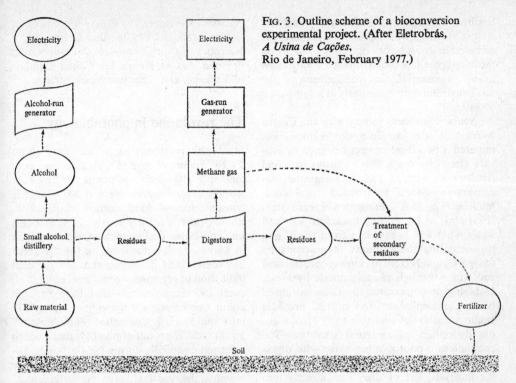

FIG. 3. Outline scheme of a bioconversion experimental project. (After Eletrobrás, *A Usina de Cações*, Rio de Janeiro, February 1977.)

microbiological method of ethanol production. Over 90 per cent of Australia's sugar production comes from Queensland. It is expected that a doubling of the crop in this state will yield sufficient ethanol to supplement between 12 to 15 per cent of Australia's fuel requirements. In the United Kingdom, the firm of Tate & Lyle 'has committed $34 million to the development of new uses for sugar and other crops as sources of energy and chemical feedstocks'. Non-conventional energy in Brazil is a primary focus of that country's energy policy. Amongst other options, the direct conversion of biomass to alcohol is a top

priority. Already a 55 kW experimental project in Cações, Bahia, generates electricity based on an integrated biomass system—producing alcohol to replace gasoline with by-product yields of gas and fertilizer.

The use of fertilizer is critically linked with energy. An essential item for agribusiness, the acquisition of high-cost fertilizers (resulting from capital-intensive technology) is outside the reach of rural producers. In Brazil, 60 per cent of fertilizers is imported, constituting a heavy drain on the national economy. As a result of low-capital bioconversion technology, plans are now under way to obtain more

117

fertilizer through a self-sufficient integrated system of energy which also yields alcohol, gas and electricity. Another option is to focus on the development of biofertilizers, or *Rhizobium* inoculant material, for management of the soil's environment, particularly in legume-crop areas.

Again, Eletrobrás jointly with the Centro Aerospacial de São José dos Campos has initiated a two-stage project designed to provide electricity (as a first priority), gas and fertilizer via alcohol and an integrated bioconversion system to isolated rural communities (Fig. 3). The integrated biogas system (IBS) aims at (a) putting back into soil and water what has been taken from them, and (b) increasing the amounts of nutrients by fixing $CO_2$ and N from the atmosphere into soil and water through photosynthesis by algae. Involving low costs managed on a decentralized basis, the implementation of IBS provides employment to the available work force without disruption of the rural structure. The coupling of a photosynthetic step with digestion also provides for the transformation of the minerals, left by digestion, directly into algae that can be used as fodder, feed for fish, or as fertilizer for increased energy production by returning the minerals to the digester process.

In addition to several other established or planned national programmes, a modest start has been made through the UNEP/Unesco-sponsored microbiological resources centre (or MIRCEN) at the Instituto de Pesquisas Agronomicas which collaborates with the Universidade Federal do Rio Grande do Sul, Porto Alegre. Already this MIRCEN has supplied 20 kg of inoculants for soybean, to Bolivia, within the framework of an FAO/United Nations Development Programme project. A similar MIRCEN at the University of Nairobi will promote analogous co-operation between Malawi, Kenya, the United Republic of Tanzania and other nearby countries.

# The revolution in photobiology

Plant photosynthesis each year fixes about $2 \times 10^{11}$ tonnes of carbon with an energy content of $3 \times 10^{21}$ joules, which is ten times the world's annual energy use and 200 times the consumption of food energy (Hall, 1979). Photosynthesis is the planet's oldest and best known biochemical reaction. Current interest in harnessing the phenomenon for the needs of progress and development is a significant indication of pressures—economic, ecological, energetic, technological—that have brought about our emergence 'from the Age of Oil into the Era of Controlled Photobiology'. Fossil fuels, the result of photosynthesis which occurred essentially during the Carboniferous era (about 300 million years ago), were once biomass; they are thus a regenerable feedstock.

Biomass is an important source of chemicals (glycerol, furfural, sorbitol, manitol). The dependence on biomass as a feedstock grows as the costs of petrochemical feedstocks grow. In specific cases, fermentation grain alcohol competes with synthetic alcohol made from ethylene. Meier (1974) regards urban recycling, 'nuplex' (nuclear-powered agricultural-industrial complexes) (cf. Malek, 1978), and aquaculture as the 'three successor revolutions to the green revolution'. Whereas the second system (nuplex) is beset by problems of high capital costs and safety, the other two systems are based to a great extent on renewable

TABLE 4. Underexploited plants of potential economic importance

| Plant | Botanical | | Remarks |
| | Family | Species | |
|---|---|---|---|
| Buffalo gourd | Cucurbitaceae | *Cucurbita foetidissima* | Alcohol production |
| Guayule | Compositae | *Parthenium argentatum* | Potential source of rubber and foreign-exchange earner for arid lands; compares well with *Hevea brasiliensis*, exploited in Malaysia |
| Guar | Leguminosae | *Cyanopsis tetragonoloba* | Untapped food resource; source of pharmaceuticals |
| Carnauba | Marantaceae | *Calathea lutea* | Source of low molecular hydrocarbons and natural product wax superior to synthetics |
| Buriti | Palmae | *Mauritia flexuosa* | Potential source of industrial starch |
| Ramie | Urticaceae | *Bohmeria nivea* | Potential source for petroleum-based synthetic fibres; potential cash/energy crop |
| Candelilla | Euphorbiaceae[1] | *Euphorbia antisyphilitica* | Source of low molecular hydrocarbons; potential foreign-exchange earner for developing countries |
| Jojoba | Buxaceae | *Simmondsia chinensis* | Yields oil-like product; potential substitute for whale sperm oil |
| Henequen | Amaryllidaceae | *Agave fourcroydes* | Annual fibre production from the leaves of the plant in Yucatan (Mexico) is 1,000 kg per hectare |
| Mesquite/ Algarroba | Mimosaceae | *Prosopis juliflora* | Source of livestock feed (pods) and wood |

1. Other potential latex-producing families are Asclepidaceae (milkweed) and Apocynaceae (dogbene).

resources, i.e. biomass, inclusive of energy crops and native plants.

### Photosynthetic energy 'factories'

Native plants, hitherto often underexploited (National Academy of Sciences, 1975), are neglected economic assets that are of enormous potential in the food, fuel, fibre and other industrial sectors pertinent to overall human societal growth (Table 4). Examples of native plants that have long been used as food are: the seeds of *Chenopodium* eaten by South American Indians, the seeds of the grasses *Sporobolus* and *Oryzopsis* eaten by North American Indians, and eel grass (*Zoster marina*) eaten by the Seri Indians. On an annual basis, the processing of *Candelilla* results in a yield of approximately 100,000 tonnes of cellulosic residue. Following fibre-extraction processes, significant quantities of residue are obtained from ixtle or *Agave lecheguilla*.

The National Council for Science and Technology and the National Commission for Arid Land Studies of Mexico have developed a project whereby 5,000 tonnes of guayule rubber obtained from *Parthenium argentatum* will be produced yearly. The process is expected to yield 40,000 tonnes of waste bagasse which will result, via the fermentation treatment process, in a potential source of bioenergy and livestock feed.

## Hydrogen—a new fuel

The production of hydrogen represents a potential alternative source of energy, as this element can be supplied to a fuel cell together with atmospheric oxygen in order to provide electricity.

The attributes of hydrogen as a fuel are as follows:

There are several methods of *formation*—chemical, electrochemical and biological.

It is *compatible* with all the primary energy sources.

It has a *good energy yield*—three times that of natural gas on a weight basis.

It can be *easily stored*—as gas, liquid or metal hydrides.

*Safety* is not a major problem; it is less toxic than other fuels.

It can be *piped*, possibly in existing natural gas pipelines.

It is ideal for *heating and cooking*—it can be used in existing burners with only minor modification required for the more intense flame.

It is usable in *internal-combustion engines* without the need for major modification of existing technology.

Hydrogen-oxygen *fuel cells* have an efficiency of over 50 per cent.

It is used in *industry* for reductive chemical processes.

It is *environmentally* very attractive—the major combustion product is water.

Biological hydrogen is possibly suitable for *delocalized production and use*, for example in remote areas and developing countries —it can be compared with methane (Smith, 1979).

R&D activities are being intensified to produce fuel cells for commercial use during the 1980s (Veziroglu, 1979; Johansson and Steen, 1977). Hydrogen utilization has actually grown with the recent use of oxy-hydrogen fuel cells in satellites, where the burning of hydrogen serves to produce electric power (see also *Mysterious Island*, by Jules Verne).

Hydrogen is produced by a large number of micro-organisms, both photosynthetic and nonphotosynthetic; Smith has discussed those microbiological processes that appear to have the greatest potential as sources of hydrogen for fuel (Smith, 1979). Recyclable, the lightest and the least polluting of all fuels, hydrogen appears to be the cheapest synthetic fuel to manufacture .Recently, a purple bacterium, *Halobacterium halobium*, has been described in which the purple pigment bacteriorhodopsin functions as a light-driven proton pump (Schreckenbech, 1976; Lanyi, 1978). Since the protons are electrically charged, this bacterium in effect converts solar energy into electrical energy. Another advantage is its high resistance to salt environments, suggesting future use in desalination processes.

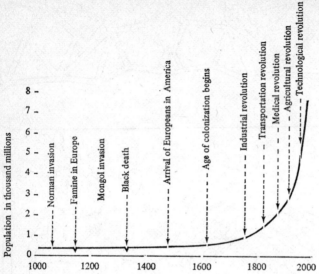

FIG. 4. World population increases with human advances in science and technology.

FIG. 5. Energy consumption increases and changes in use pattern with human advances in science and technology. (After E. Taiganides, Energy and Useful By-Product Recovery from Animal Wastes, in E. Okano, B. Iohani and N. Thanh (eds.), *Water Pollution Control in Developing Countries*, Bangkok, Asian Institute of Technology, 1978.)

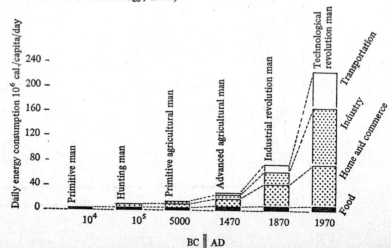

# Conclusion

The discovery and processing of fossil fuels have been channelled essentially towards meeting the exponential growth needs of the population and its technological accomplishments that are heavily dependent upon energy (Figs. 4, 5). Increasing demographic and industrial burdens have bared both the limited availability of fossil resources and the necessity of developing alternative energy sources.

The OPEC countries are in a position to subsidize their domestic agricultural programmes from proceeds on the sales of petroleum. Both developed and developing countries, endowed with other assets like bauxite or copper, are perhaps able to soften the impact of high fossil-fuel costs through their own foreign-exchange earnings. In the final analysis, it is many of the developing countries, with their huge population and traditional agrarian economies, that have the most pressing needs for fuel in the domestic, agrarian and industrial sectors.

In the non-OPEC developing countries (inhabited by more than 40 per cent of the world's population), 90 per cent of their total energy use is derived from non-commercial fuel—wood, dung and agricultural waste. Current practices in most countries, in both rural and village settings, involve an excessive use of wood fuel; this, in turn, leads to deforestation, depletion of other valuable resources, causation of ecological imbalances, and ultimate climatic change. In counteracting deforestation and desertification, the use and importance of several agricultural residues such as bagasse and straw, plants like reeds and grasses, and non-wood crop fibres (e.g. hen-

equen), have been discussed in terms of the pulp and paper industry.

In the return to 'sense from non-sense', the new approach to agriculture is characterized by a minimum of waste of energy, minerals, the reduced use of fertilizers and pesticides, a much larger diversity of crops, the use of aquaculture, and, most important of all, the utilization of biotechnology on household, village and industrial scales. Thus the recycling of agricultural residues—either through fermentation or bioconversion processes, which do not involve fuel consumption—can help meet the energy requirements of the developed and especially the developing countries.

# References

BULL, A.; ELLWOOD, D.; RATLEDGE, C. 1979. The Changing Scene in Microbial Technology, *Microbial Technology, Society for General Microbiology Symposium 29*, pp. 1–27.

DASILVA, E. 1978. Bio-Gas Generation: Developments, Problems and Tasks (An Overview). *Proceedings of the Joint WH-NR United Nations University Conference, Bioconversion of Organic Residues for Rural Communities*. Guatemala, November.

DASILVA, E.; OLEMBO, R.; BURGERS, A. 1978. Integrated Microbial Technology for Developing Countries: Springboard for Economic Progress. *Impact of Science on Society*, Vol. 28, No. 2, pp. 159–82.

GRUBER, E. 1978. Making Use of Biomass: Is This the Long-term Solution to Energy, Food and Raw Material Problems? *Universitas*, Vol. 20, pp. 113–22.

HALL, D. 1979. Plants as an Energy Source. *Nature*, Vol. 278, pp. 114–17.

JOHANSSON, T.; STEEN, P. 1977. *Solar Sweden—an Outline to a Renewable Energy System*, p. 109. Stockholm, Secretariat for Future Studies.

LA RIVIÈRE, J. 1978. Environmental Goals for
Microbial Bioconversion in Rural
Communities. *Proceedings of the Joint
WH-NR United Nations University
Conference, Bioconversion of Organic Residues
for Rural Communities*, Guatemala,
November.

LANYI, J. 1978. Light-Energy Conversion
in *Halobacterium halobium. Microbiological
Reviews*, Vol. 42, pp. 682–706.

LIPINSKY, E. 1978. Fuels from Biomass:
Integration with Food and Materials System.
*Science*, Vol. 199, pp. 644–51.

MALEK, I. 1978. Production of Protein in the
Agro-Industrial Complex of Nuclear
Energy Centers Connected with the
Desalination of Seawater. In: W. Stanton
and E. DaSilva (eds.), *State of the Art
of Applied Microbiology in Relation to Its
Significance to Developing Countries*.
Kuala Lumpur, University of Malaya Press.

MEIER, R. 1974. *Planning for an Urban World*.
Cambridge, Mass., MIT Press.

NATIONAL ACADEMY OF SCIENCES. 1975.
*Under-exploited Tropical Plants with Promising
Economic Value*. Washington, D.C.,
National Academy of Sciences.

——. 1977. *Methane Generation from Human,
Animal and Agricultural Wastes*.
Washington, D.C.,
National Academy of Sciences.

PORTER, J. 1978*a*. Microorganisms as Tools for
Rural Processing of Organic Residues.
*Proceedings of the Joint WH-NR United
Nations University Conference, Bioconversion
of Organic Residues for Rural Communities*,
Guatemala, November.

——. 1978*b*. Microbiology and Biomass
Agriculture. Paper presented in *Symposium 16:
Production of Agriculturally Useful
Products by Microbial Conversion at the
XIIth International Congress of Microbiology,
Munich, September 1978*.

SCHRECKENBECH, T. 1976. Light-Energy
Conversion by the Purple Membrane from
*Halobacterium halobium*. In: H. Schlegel
and J. Barnea (eds.), *Microbial Energy
Conversion*. Göttingen, Erich Zoltze.

SMITH, G. 1979. Microbiological Hydrogen
Formation. *Search*, Vol. 9, pp. 209–13.

UNITED NATIONS. 1978. *Report of the Group
of Experts Meeting on Waste
Recycling Technology for Development,
11–15 December 1978*. New York, Office of
Science and Technology, United Nations.
(Document IESA/S+T/AC.11/15.)

VEZIROGLU, T. 1979. An Energy Infrastructure:
Hydrogen Energy System. *Interciencia*,
Vol. 4, pp. 32–7.

Edgar J. DaSilva

# Energy from the wind

## 12

United Nations Environment Programme

*The wind arises from atmospheric temperature differences created by the sun. It has propelled
ships at sea and powered (via windmills) grinding, pumping and other operations for thousands
of years. More recent interest in harnessing the wind, the technology of which is essentially
in hand, sees increased use for water pumping and in rural electric systems. Some quite novel uses,
such as compressing air to aerate ponds, are also under active development.*

Wind energy is a renewable source of energy
manifested by the power of the sun. The source
of wind is in the atmospheric temperature
differences generated by the sun, which in turn
give rise to pressure differentials. The wind is
a mechanism for dissipating, as kinetic energy,
the potential energy accumulated in those
pressure differences.

Wind energy has been used for thousands
of years to propel boats and ships and to provide mechanical power for grinding grain and
processing other agricultural products. The
significance of wind power peaked in the
sixteenth century, and in 1850 windmills provided about $1 \times 10^9$ kW (McGowan and Heronemus, 1975). By the turn of the century there
were more than 30,000 windmills in Denmark
and more or less similar numbers in the
Netherlands and other countries, used mainly
for grinding and for pumping water.

Edited version of Chapter IV: 'Wind Energy',
UNEP Report, 1980.

Different authors have estimated the global
wind-energy resource base to be in the range
of 1,200 TW (Sørensen, 1979), but the harnessing of this power is very site dependent.
Average wind velocity, at a height of 20–30 m
above ground level, should be high enough so
that the energy flux (that is, the 'power' in the
wind) through a properly oriented vertical
section would reach an annual average suitable
for conversion. A wind machine located at a
site where the annual average wind-power is in
the range of 500 W/m² (7 m/sec wind speed)
can convert to electricity about 175 of the
500 W.

The energy flux contained in a moving airstream is proportional to the cube of the wind
velocity. Not all the energy in the wind can,
however, be extracted, even by an ideal
machine. Theoretically, the maximum fraction
of the power in the wind that is extractable
is 59.3 per cent (Merriam, 1978). In practice,
the best power-extraction efficiency reported
for a real machine is about 0.50, and this is not

achieved at all windspeeds, but only at the single windspeed for which the wind machine design has been optimized. Also, the conversion of mechanical to electrical energy takes place at less than 100 per cent efficiency; typically 75–95 per cent. Adding all factors together, the electrical power delivered by a real wind generator is perhaps 30–40 per cent of the power in the wind when the wind generator is operating smoothly at some windspeed within the design range. However, windspeeds occur that are not within the design range. There is a windspeed below which the machine will not operate at all ('cut-in speed') and there may also be one above which the machine is designed to feather, turn out of the wind or otherwise protect itself ('furling speed').

If the windspeed exceeds rated windspeed some of the extractable power in the wind is wasted in order not to exceed the electrical rating of the generator. Considering these factors the electrical output over a year's time might be 15–30 per cent of the energy in the wind, or even less, depending on the site and the characteristics of the machine.

The technology of wind machines, and also their history, is exhaustively discussed in the literature (see, for example, Torrey, 1976; Simmons, 1975; Merriam, 1978). The physics and engineering of the machines is well understood. Testing and modification are in progress and mass production may be established in the near future. The very large wind generators to date have been prototypes. All have operated for limited periods. Failure or shutdown has occurred; however, this could have probably been avoided with proper engineering work. There are no substantial technical problems that would limit large-scale development of wind machines in the

near future. The dominant technology in most national programmes today is the two-blade propeller turning about a horizontal axis. The vast majority of large wind-electric generators constructed in the past have been two- or three-blade propellers. Other technologies, such as the vertical-axis Darrieus rotor, single-blade, horizontal-axis rotors, and others, may offer advantages, but the advantages are not likely to be crucial for the development of wind energy.

In spite of the lack of extensive wind data and problems of site selection, it is possible to make gross worldwide estimates of possible wind contribution to the energy budget. The crucial factors that will determine whether or not wind energy contributes in a substantial way to human energy needs over the next half-century are technological in nature, and relate mainly to national energy policies, cost and public acceptability. Table 1 gives an estimation of the possible contribution of wind energy to world energy supplies until the year 2000.

The potential wind power extraction along continental shore-lines is shown in Table 2.

TABLE 1. Estimated possible contribution of wind to world energy supplies

| Year | Installed capacity (kW) | Energy output kWh/year |
|---|---|---|
| 1985 | $0.5 \times 10^6$ | $0.15 \times 10^{10}$ |
| 1990 | $10 \times 10^6$ | $3 \times 10^{10}$ |
| 2000 | $200 \times 10^6$ | $90 \times 10^{10}$ |

*Source:* Merriam (1978). For comparison, the total electricity production in the world in 1976 was $760 \times 10^{10}$ kWh (United Nations, 1979).

TABLE 2. Estimate of coastline wind potentials

| Region | Assumed coastline ($10^6$ m) | Estimated coastline wind potential (TWh per year) | Electricity consumption (1978) (TWh per year) |
|---|---|---|---|
| North America | 46 | 754 | 2 400 |
| South America | 22 | 604 | 300 |
| Oceania | 20 | 780 | 150 |
| USSR | 11 | 494 | 1 000 |
| Asia (excluding USSR) | 42 | 701 | 1 000 |
| Europe (excluding USSR) | 24 | 1 051 | 2 000 |
| Africa | 27 | 534 | 200 |

*Source:* Sørensen, 1979.

The most probable trend in the development of wind energy in the near future is a much increased use of wind machines in the 5 to 100 kW range for water pumping and rural electrical systems with two-day storage batteries. Machines for generating electrical power into existing utility grids in the range of 100 kW to 5 MW are also being developed.

In addition to the classical applications of wind-power a number of other future applications has been considered. These are briefly described in the following:

The production of fertilizer using wind-power: static electricity produced by a windmill system can be sparked in an airstream and the oxides of nitrogen thus produced are absorbed in water forming a nitric-acid solution. Prototypes of this system have been investigated. Given the large demand for nitrogen-based fertilizers in the world, the possibility of developing initially small-scale systems to produce fertilizers in remote areas, islands and mountainous regions could reduce the transportation costs.

The generation of electricity through wind-power can be applied to existing electrolysis technology for the production of hydrogen and oxygen.

The use of windmills to compress air, which can then be bubbled at the bottom of river-beds and harbours to prevent the build-up of ice, has been discussed as a future possibility in cold regions.

The use of wind-power to compress air which can then be used for the aeration of various bodies of water, particularly in warm regions and during summer periods, can be envisaged to overcome the problems of stagnation and depleted oxygen in bodies of water. This has an application in aerating fish-ponds and other water basins where there are detrimental environmental effects. Several commercial firms are already marketing small wind-operated air compressors feeding aeration systems placed in ponds.

christie

Several investigations have been proposed and studies for the use of wind-power to operate wind-powered desalination systems for converting brackish water or sea-water into fresh water. Generally the wind-power would be used to provide mechanical shaft-power to operate modified reverse osmosis or vapour-compression desalination plants.

Given the fact that there are considerable wind-power potentials off the coastline of a number of countries (see Table 2), the problem of generating electricity through the use of windmills located on off-shore platforms has been given some attention and will no doubt be considered more seriously in the future. The electricity that is generated can be transmitted through underwater cables and coupled with the national grid on shore. The technologies of platform construction, used in off-shore drilling activities in the petroleum field, as well as the transmission of electric power through underwater cables are already established.

## References

McGowan, J. G.; Heronemus, W. E. 1975. Ocean Thermal and Wind Power. *Environmental Affairs*, Vol. 4, No. 4, p. 629.

Merriam, M. F. 1978. Wind, Waves and Tides. *Annual Review of Energy*, Vol. 3, p. 29. Palo Alto, Calif., Annual Reviews Inc.

Simmons, D. M. 1975. *Wind Power*. Park Ridge, N.J., Noyes Data Corporation.

Sørensen, B. 1979. *Wind Energy*. UNITAR Conference on Long-term Energy Sources. New York, United Nations.

Torrey, V. 1976. *Wind-catchers*. Brattleboro, Vt., Stephen Greene Press.

United Nations. 1979. *World Energy Supplies 1973–1978*. New York, United Nations. (Statistical Paper, J22.)

# Solar energy in developing countries    13

A. Ramachandran and J. Gururaja

*Mankind's progress is profoundly influenced by the discovery and exploitation of different sources of energy. The use of coal, the development of hydroelectricity, the discovery of oil and gas, and the advent of nuclear energy are significant milestones in history. Each new source heralds the emergence of a range of new technologies; solar energy now figures as one of these 'new' sources.*

The use of energy is an integral part of the development process—no matter how one defines development, whether in terms of increased gross national product or of human welfare. Of the world's 4,000 million people, 1,000 million in North America and Europe use 84 per cent of the energy supply; 2,000 million in Brazil, China, India and a few other countries consume 15 per cent, while the remaining 1,000 million in Africa, Asia and Latin America use the remaining 1 per cent. The developing countries generally have not only a very low rate of energy consumption per capita, they also exhibit within their boundaries a strikingly inequitable distribution of energy between rural and urban populations.

Until recently the world had been complacent about the energy supply and had hardly given thought to long-range problems of the availability of fossil fuel and to technologies for exploiting new and renewable sources of energy. Petroleum still occupies a dominating position, but indications are that its availability will taper off in the beginning of the next century. The developed and developing countries have realized the dire need to diversify energy sources and place greater reliance on renewable sources of energy. Among the several potential new sources of energy, solar energy holds promise as a major supplementary source. Worldwide attention is now focused on methods of exploiting solar energy for a vast range of applications.

Discussion of potential application of solar energy and its possible benefits (by way of increasing energy supply in rural areas and in reducing human drudgery) will require consideration of sunlight's direct and indirect forms and its associated technologies. One should recognize that traditional sources of

Originally published as 'The Prospects of Solar Energy for Developing Countries', *Impact of Science on Society*, Vol. 29, No. 4, 1979, pp. 319–26.

energy in most developing countries, especially in rural areas, are already based heavily on the indirect conversion of solar energy; this occurs through photosynthesis in the form of fuel wood and agricultural residues, which constitute a major portion of the total energy consumption. Indeed, biological applications of solar energy are gaining increasing importance. Efforts aimed at attaining a sustainable rate of consumption of these sources, through planned use of forests and development of fast-growing species of plants, are considered essential.

The use of solar energy in the developing world must be viewed in different countries within the national context; they must take into account local rural situations. The foremost requirement is an inventory of energy needs, identifying the relative quantity of energy required for various activities and, in addition, the pattern of energy consumption relating this quantity to the nature of its end-use. Thus it is considered extremely important that new energy technologies should be able not only to provide the required amount of energy but that this should be matched to the quality of its end-use.

Here we discuss various energy needs that could be met by using solar energy, keeping in mind the element of appropriateness to the situations in developing countries. The energy is required for domestic, agricultural, household and village industries, and community purposes. While it is difficult to envisage that any of the new and renewable sources of energy would be able to offer short-term solutions on a significant scale, they do offer the prospect of indigenous, readily available, and inexhaustible supplies which—through sustained efforts—can bring about an improvement in the quality of life in rural areas. A further advantage of renewable energy sources is their compatibility with the environment.

## Energy needs in developing countries

Study of energy use in rural areas of developing countries have revealed that cooking and agricultural operations make the heaviest demands on energy budgets and labour. The paucity of reliable data has led to differences amongst the various studies. For example, Roger Revelle found that cooking accounts for 60 per cent and agriculture for 22 per cent of the total energy consumption of 7,100 kcal/capita/day (which is 380 kg in coal equivalent/year).[1] Makhijani and Poole calculated that agricultural and cooking needs absorb from 77 per cent to 99 per cent of the total per capita energy consumption,[2] which ranges from 670 to 2,260 kg in coal equivalent (ce), and Reddy and Prasad estimate that 130 kgce is used for cooking out of a total per capita of 300 kgce.[3] Against these estimates, J. Parikh has developed the figure of 300 kgce for cooking alone.[4] These differences notwithstanding, energy consumption per capita varies tremendously among developing country villages, both in volume and mix.

1. R. Revelle, 'Energy Use in Rural India', *Science*, Vol. 192, 4 June 1976, p. 969.
2. A. Makhijani and A. Poole, *Energy and Agriculture in the Third World*, Cambridge, Mass., Ballinger, 1975.
3. A. Reddy and K. Prasad, 'Technological Alternatives and the Indian Energy Crisis', *Economic and Political Weekly*, Special Number, August 1977, pp. 1465–1502.
4. J. Parikh, unpublished work.

TABLE 1. Principal energy needs in rural areas of developing countries

| Modes of energy use | Specific energy needs | Modes of energy use | Specific energy needs |
|---|---|---|---|
| Domestic | Lighting<br>Cooking<br>Hot water<br>Heating | Agricultural processes | Grinding<br>Drying<br>Storage<br>Refrigeration |
| Water pumping | Drinking and household<br>Cattle watering<br>Irrigation | Rural industry | Pottery<br>Spinning and weaving<br>Furniture |
| Agriculture | Irrigation<br>Planting<br>Cultivation<br>Harvesting | Metalwork<br>Transport | Metalwork<br>On land<br>On water |
| | | Social and community | Health services<br>Drinking water treatment<br>Education |

Even though cooking requires a dominant share of energy, the possibility of finding substitutes for wood and dung used for this purpose are limited. Fuel is used very inefficiently for cooking; the efficiency of utilization ranges from 5 to 10 per cent. The improvement of efficiency of wood-burning stoves, with perhaps marginal increase in cost, should be a priority area for research.

The major energy needs in rural areas of developing countries are listed in Table 1.

## Potential uses of direct solar energy

### Cooking

Fuel wood, agricultural and forest residues, dung cakes and (to a limited extent) kerosene are currently the combustibles used for cooking in rural areas; these are obtained at essentially no private cost to the individual. The adverse impact of wide use of these materials on the environment is now widely recognized. Whether or not solar cookers could provide a new solution is moot and does not depend solely on the technological aspects, but rather on social habits and acceptability. Technologically there is a wide variety of designs. The functional requirements of a solar cooker are that the solar heat should (a) be concentrated to temperatures of about 300 °C, suitable for use inside a house, and (b) have the necessary storage to deliver heat for even a few hours after sunset. The currently available concentrator-type designs or hot boxes do not satisfy these functional requirements and are either too expensive or functionally unacceptable to rural populations. Further research on heat-storage materials and chemical salts may lead to acceptable solutions but, for near-term application, fuel wood may continue to be important. Biogas for cooking will increasingly become prominent in many parts of the developing world.

## Water pumping

Solar-powered water-pumping systems could be based on conversion of solar energy to mechanical energy to be used directly, for pumping, or converted to electricity. In the latter case, electrical energy can provide not only for pumping but also for industrial and domestic uses. In the former case, the pumps could be located on dispersed farms without requiring an electrical grid. The use of small photovoltaic pumps for drinking water and micro-irrigation is another promising technology. In a number of developing countries, the perfecting of solar pumps has been accorded a high priority.

Water requirements for irrigation depend on climate, soil and type of crop. Daily water requirements for sugar cane grown in a twelve-month cycle, for example, in India, vary by a factor of five during the year, ranging from 2.8 mm/day to 14.3 mm/day (2,000 mm/annum). *Jowar*, or millet, is grown in a four-month cycle, with the possibility of double cropping. It is the availability of water in the third fortnight after planting that is of importance. Rice is usually grown in permanent standing water. This makes possible a continuous water supply by pumping during the fourth cycle (100–120 days), as the field itself serves as storage. During transplantation, however, the whole land area has to be inundated, and the water supply needs to be four to five times the average.

Owing to the requirement for the proper

quantity of irrigation water at the proper time in the agricultural operations of a given crop, pumping equipment used in irrigation usually has a low load factor. Therefore care is required in defining the allowed maximal cost of irrigation water and then using this as the basis for comparisons with solar devices. Solar pumping costs would be high because of the high capital cost of equipment. It is thus necessary to aim at a high load factor for the basic system; for this purpose, it is essential that multiple uses of energy conversion system be built in during the design stage.

Pumping systems can be based on flat-plate or focusing collectors. The main problem in using a flat-plate system is its relatively low efficiency—with engines running at temperatures of less than 100 °C, requiring large collector areas per unit of mechanical energy. The capital cost of such systems, judged from the type of pumps already installed in Africa by one French company, is prohibitive. Its concentrator system would have higher efficiency, but the system becomes too complicated for rural applications. One solution is to build a completely sealed unit requiring minimal maintenance, but current efforts to do so remain largely unsuccessful. Under certain circumstances, photovoltaic pumps are nearly competitive with diesel-powered pumps, whose costs are usually higher than with electric pumps. If the cost of solar pumping, based on a projected price of $1 per peak watt (the near-term cost goal), is compared with the current cost of diesel pumping, the photovoltaic pump option is about 25 per cent more expensive than diesel pumping, as shown in Table 2.

These figures should be looked at with much caution, since the calculation is based

TABLE 2. Comparison of diesel *v.* solar-powered irrigation pumps in India (in rupees). Capacity: 50 kilolitres/day at 5 to 8 metres of head for 300 working days; assumptions: (i) cost of solar cells is 10 rupees (approximately \$1) per weak watt, as expected by 1985, (ii) cost of diesel oil is 1.5 rupees/litre

| Solar water supply | Rupees | Diesel water supply | Rupees |
|---|---|---|---|
| *Capital cost* | | | |
| 1-kW solar panel | 10 000 | Diesel engine | 3 000 |
| Battery wiring, switches, etc. | 4 000 | Pump set, boring, pumping, etc. | 3 000 |
| 500-watt motor pump set, boring well, pumping, etc. | 3 500 | | |
| Total investment | 17 500 | | 6 000 |
| | | | |
| *Annual cost* | | | |
| Depreciation of solar panel (20 years) | 500 | Depreciation (10 years) | 600 |
| Depreciation of other equipment (10 years) | 750 | 10 per cent interest | 600 |
| 10 per cent interest | 1 750 | Diesel fuel at 0.5 litre/h, 300 working days | 1 125 |
| Maintenance | 500 | Maintenance | 500 |
| Total annual cost | 3 500[1] | | 2 825[2] |

1. Cost of water=0.24 rupees/kilolitre.
2. Cost of water=0.19 rupees/kilolitre.
*Source:* Agarwal, 'Solar Energy and the Third World', *New Scientist*, 9 February 1978, p. 358.

on approximate values and simplified assumptions.

If diesel fuel prices increase further as expected, solar photovoltaic pumps may well become very attractive.

**Solar electricity**

The conversion of solar energy to electricity, either by thermal energy conversion or direct conversion, is of interest to developing countries inasmuch as the benefits of rural electrification have reached only a fraction of all villages; its implementation on a wider scale involves heavy capital investment which the communities could ill afford. Furthermore, the high transmission and distribution losses, together with the high cost of carrying electricity to remote areas, have led to the consideration of decentralized electricity generation in villages (as opposed to central generation and transmission). Although in many developing countries diesel generators are being used in remote areas, the rising cost of such generation and the burden of imports point to the need to explore alternative systems.

In India, for example, though rural electrification has made significant strides, only about 40 per cent of a total of 5,600,000 villages have been electrified, and the bulk of these are villages close to cities or the power grid. The rural community is highly dispersed; as a matter of fact, nearly 60 per cent of the villages have a population of less than 500. It is for such villages that the introduction of solar power plants would be appropriate and justifiable, provided that the costs are competitive. Cost-benefit analyses are somewhat meagre, and even those that have been made in regard to the viability of solar technologies do not always take into account the social and environmental cost advantages of these technologies. Although it may be difficult to quantify them, these advantages are considered important from the long-term point of view.

Turning to the cost of solar electricity, it is determined by the size and type of plant, load factor, service life, and a host of other elements. For small dispersed communities, it is thought that the capacity could be in the range of 10 to 100 kW. Solar power-plants based on flat-plate collectors have an inherently low efficiency, while concentrating systems become rather complex for village use. There is inadequate field experience with both the technical and non-technical aspects of solar power-plants, although a few plants have indeed been installed on an experimental basis, including 10-kW flat-plate systems in India and Egypt. Solar power plants in the range of 1,000 kW and higher—though being investigated in the United States and Europe—are not particularly suitable for developing countries in the near term (0–5 years) or medium-term (5–10 years).

The current cost of solar thermal electric systems quoted is around $8,000 to $10,000 per kW for systems of 10 to 100 kW. As for other large systems, such as solar towers, dish systems and the like, the cost goal is $4,000 per kW. We may expect, therefore, that the cost of solar thermal electric systems of small capacity will be about $8,000 in the near term, with the possibility of the cost diminishing to $4,000 in the mid-1980s. There is the possibility that the cost will be half of this in the early 1990s. Whether solar power-plants would eventually be competitive or not would depend largely on the increase of cost of conventional fuels, the environmental costs of conventional power production, and the resolution of many technical problems of such systems. There is, however, sufficient ground to justify enhanced research and development efforts in this field.

Photovoltaic conversion of solar energy directly to electricity holds great promise for small-scale application in rural areas and remote locations. Rapid developments are taking place not only in the technology of monocrystalline silicon solar cells, but also in other promising fields: polycrystalline and amorphous materials, and thin-film cells. Costs have fallen from $100,000 per kW in the 1950s to about $10,000 per kW today. The cost goal generally accepted by the industry is $500 per kW (peak), which is expected to be attained by 1986–87. At these prices photovoltaic systems will prove to be attractive for a wide variety of applications in developing countries, such as running small electric pumps for rural drinking-water supply, local source of electricity for lighting, community radio, television

Christé

and other communication purposes. The main breakthrough will have to come from low-cost materials and processes for the production of cells and through mass production techniques. Equally important is a programme of field testing and evaluation of photovoltaic systems in order to establish their suitability for village applications.

## Solar crop-drying

One of the most compatible systems for agricultural use is a solar hot-air system suitable for crop drying. Post-harvest technology aimed at improving the quality and reducing the losses of food grains is a crucial aspect of the food problem in developing countries. Solar drying, as a viable technology, has now been demonstrated for applications in fruit drying, fish drying, and the drying of general agricultural produce such as timber, grain, cash crops like ginger, cashew, pepper and coffee. A number of development projects are currently in progress for use in non-industrialized countries.

Small-scale drying technology is relatively well established, although its commercial availability is limited. Large-scale systems are gaining acceptance. Expected improvements in technology, together with promotional efforts in rural areas, are likely to result in applications of solar dryers on a wider scale.

## Water-heating systems

Solar hot-water systems for domestic use and for space-heating purposes are now well-established technology. Except in northern latitudes where the winter months are long, the potential for hot-water systems in developing countries is somewhat limited. But it is now becoming increasingly evident that, for many agriculture and village industries (such as handloom fabrics, leather tannery and hand-made paper), the potential for solar hot-water systems is considerable and worth pursuing seriously. Although the technology is reasonably well developed, further cost reductions are necessary.

## Desalination

Solar distillation to convert brackish or sea water to potable water is used on a small commercial scale in some countries. The technology is simple, and small solar stills can be fabricated locally in rural areas. Efforts to improve performance and reduce costs are worth while, as this is an area of considerable importance to developing countries.

## Cold storage of perishable foodstuffs

Solar refrigeration units employing the refrigeration principle of absorption have been demonstrated in various sizes. In spite of the great importance attached to this application, especially in view of the useful role cold storage could play in the preservation of perishable articles, the technologies developed are not yet readily available on an economical scale. Estimates of economically feasible capacities range from 5,000 to 20,000 kcal/day corresponding to ice production of 40–200 kg/day, adequate for the needs of a sizeable community. Further research and development would be desirable in this area.

# Potential for indirect use of solar energy

## Wind energy

The potential use of wind energy in developing countries is considerable, although highly location-specific. For example, in most parts of India the average wind velocity is low, requiring windmills that could operate at low wind speeds. Where average wind velocities are sufficiently high, windmills for pumping and electricity generation would prove attractive.

Wind-powered electric generators have been produced for a number of years on a small scale, and the cost of such units is around $2,000/kW, but if these are made in large numbers the cost might descend to about $1,000/kW. Attempts are under way to develop inexpensive windmills based on local materials and using local skills. Emphasis here is not so much on service life as on simplicity, low cost, and ease of repair and maintenance with local resources. As with other solar systems, storage is an important requirement in wind-powered electric systems, whereas if the windmill is directly used for pumping water, storage is not critical. The wind-powered electric option for villages is highly location-specific and further development is required to establish the viability of wind energy systems in developing countries having different agricultural and climatic conditions.

## Bioconversion

Biogas produced by the anaerobic digestion of organic wastes, as well as human wastes, holds considerable promise for developing countries. There are many advantages to be drawn from the use of biogas, especially in rural areas. It is reported that China already has commissioned a few million biogas units of various sizes and designs. Other countries, notably the Republic of Korea and India, have been trying to introduce biogas units in rural areas in a major way. In some regions, dried cattle dung has been traditionally used as domestic fuel—robbing the soil of an important fertilizer, thus reducing productivity. The use of biogas units provides a sanitary and convenient way to produce fuel while retaining a large part of the feedstock in the form of sludge, which can then be used as a very effective soil conditioner and fertilizer.

While many units are in operation, experience to date shows that capital costs of a digester are still high and beyond the reach of the average rural family in many developing countries. Further work in this field is required to arrive at the best possible design of a digester that gives an optimal yield of methane gas and that is insensitive to operating conditions and involves little maintenance. A multidisciplinary group would be necessary to look into not only the design aspects but such factors as integrating the development of biomass into overall rural development strategy. Community size units, as opposed to family size units, would be more advantageous to the poorer sections of the rural population. Necessary information with regard to the best sizes for community plants and their operational problems is not yet available.

### Fuel crops for energy

The renewable nature of the forest offers potential for a sustained output of wood for fuel, provided recycling can be ensured by the appropriate harvesting and management of forests. The energy of photosynthesis is fixed in tree biomass and can be released in a number of ways. In addition to direct burning, wood can be processed to produce gaseous or liquid fuels as well as charcoal. Charcoal technology is well known, involving wood distillation, pyrolysis and heating in the absence of oxygen to produce, together with charcoal, gaseous by-products which can be used either directly as fuel, or can be distilled to produce acetic acid, methyl alcohol and other liquid substances. Ethanol production through the fermentation of biomass also offers distinct possibilities. Although the technology is rapidly developing, a great deal of further research and development are still needed. No doubt, new policy measures on the part of government and participation of local people would be crucial in this expansion.

## Integrating direct and indirect solar energy

The experience of the last few years has shown that a system approach to energy problems is likely to optimize benefits obtainable from renewable energy sources. Whatever new energy sources are available in a given location, their integration in such a way as to be able to meet diverse energy requirements is indeed a challenging task. Most research on alternative energy sources has concentrated on single sources like solar, or biogas or wind or geo-

thermal power, and often on a single application. The intermittent nature of some sources, like the sun or the wind, is one of the major problems in their effective utilization. Combining several sources would help to alleviate this problem and improve system reliability. Unlike sun and wind, biogas could provide a continuous source—provided, of course, that feedstock is maintained continuously at the required level. Integrated systems based on sun, wind and biogas are worth pursuing in developing countries. The scope of integrated systems should be most attractive to villages. But it is necessary to keep in mind that operational skills in villages are usually poor and, for this reason, technology should be simple and maintenance requirements low.

The need to investigate the feasibility of integrating various alternative energy sources ought to be seen from two perspectives. First, integration would increase the efficiency of an energy source, e.g. biogas yield is increased by heating the digester. The heat source could be supplied by the sun. Secondly, integration provides a more reliable energy supply, e.g. multi-source input to electricity generation and solar heating, boosted by biogas.

It is to be noted that there are significant gaps in current knowledge of the status of the energy situation in developing countries and of the range of various solar technologies that will meet energy needs, in respect to both quality and quantity. The lack of adequate information regarding the availability of suitable materials, manufacturing capabilities, and skills in the less industrialized countries render them inappropriate for use or unsuitable for manufacture in developing countries. The creation of solar energy information systems in each country and mechanisms for the exchange of information is of vital importance for the promotion of the use of solar energy in developing countries.

Solar technology is still in its early days, and teething troubles are inevitable. Since it is, in financial terms, a high-risk activity, many of the resources for solar technology will have to come from government support and international co-operation.

# A question of strategy

# 1980-2030: two scenarios

## 14

Alan McDonald

*This explores in detail future energy demand and the competition among different energy sources contributing to meeting this demand. We extend the analysis only as far as 2030. The quantitative results are expressed in two reference scenarios and three supplementary cases which are variations of the reference scenarios. The principal tool used in building the scenarios and alternative cases is the set of computer models outlined briefly.*

The two scenarios are labelled 'high' and 'low'. The former assumes relatively higher economic growthrates throughout the world, and the latter assumes relatively lower worldwide economic growth. The high scenario leads to a level of global primary energy consumption in 2030 equal to 35.7 TWyr/yr, which amounts to slightly more than four times the 1975 level of 8.2 TWyr/yr, while the low scenario yields a global primary energy consumption in 2030 of 22.4 TWyr/yr, a little less than three times the 1975 level.

The two scenarios are not meant to describe extremes in either direction, but rather to cover a middle ground. Neither are they intended as predictions; instead, the objective was to detail the engineering and economic consequences that follow from two different sets of reasonable assumptions. None the less

the results of the exercise suggest powerful trends within our current global energy system, and it is worth listing these before describing the scenarios.

In the developed regions of the world there is a tremendous potential for energy conservation from efficiency improvements and expanding the economic sectors that are less energy intensive, such as the service sector. For these regions the average growth rate for final energy from 1975 to 2030 is only 1.7 per cent per year in the high scenario and 1.1 per cent per year in the low scenario. These values compare to a 1950–75 average of 3.8 per cent per year.

In the developing regions expanding populations, increasing urbanization, and continuing development needs limit the prospects for energy savings. As a result, throughout the 1975–2030 period primary energy growth rates in these regions are predominantly higher than the gross domestic product (GDP) growth rates, although

An edited version of Section 4: '1980–2030: Demand, Conservation, and Two Scenarios', (IIASA, 1981).

the differences tend to decrease with time. In contrast, in the developed regions the primary energy growth rates are always below the GDP growth rates.

The production and consumption of oil in both scenarios go up, not down, compared with 1975. Although oil's share of the primary energy market decreases from 1975 to 2030 (from 47 to 19 per cent) in the high scenario and from 47 to 22 per cent in the low scenario), the absolute amounts of oil used go up (from 3.83 TWyr/yr in 1975 to 6.83 TWyr/yr in 2030 in the high scenario and from 3.83 TWyr/yr in 1975 to 5.02 TWyr/yr in 2030 in the low scenario).

Despite such increases, and even with vigorous conservation measures in the industrialized regions, increasing needs for liquid fuels throughout the world may, over the next five decades, exceed the capabilities of the global energy supply system. In the high scenario, primary liquid-fuel demand increases from 3.83 TWyr/yr in 1975 to 11.1 TWyr/yr in 2030. In the low scenario the increase is from 3.83 TWyr/yr in 1975 to 7.22 TWyr/yr in 2030. These 2030 demand levels exceed 2030 oil production levels by 63 and 44 per cent for the high and low scenarios, respectively.

The gap between liquids demand and oil supply is closed by liquefying tremendous quantities of coal. For the high scenario, 6.7 TWyr/yr of coal are liquefied in 2030; for the low scenario the figure is 3.4 TWyr/yr. For both cases this amounts to liquefying more than half the coal mined in 2030. (The high-scenario value of 6.7 TWyr/yr of coal is equivalent to 4.3 TWyr/yr of crude oil, which nearly equals the total world crude oil production of 1978.)

Any oil produced will come increasingly from unconventional sources, such as tar sands, oil shales, heavy crudes and enhanced recovery techniques. In the high scenario the shift is such that by 2030 the majority of the oil produced is, in fact, unconventional oil.

The forces driving energy demand can be divided into four categories: (a) population growth; (b) economic growth; (c) technological progress; and (d) structural changes within economies.

TABLE 1. Global population projections by region (in millions)

| Region | Base year 1975 | Projection | |
|---|---|---|---|
| | | 2000 | 2030 |
| I (NA) | 237 | 284 | 315 |
| II (SU/EE) | 363 | 436 | 480 |
| III (WE/JANZ) | 560 | 680 | 767 |
| IV (LA) | 319 | 575 | 797 |
| V (Af/SEA) | 1 422 | 2 528 | 3 550 |
| VI (ME/NAf) | 133 | 247 | 353 |
| VII (C/CPA) | 912 | 1 330 | 1 714 |
| World | 3 946 | 6 080 | 7 976 |

*Population growth.* The assumptions about population growth are presented by region[1] in Table 1. We see that 90 per cent of the projected population growth between 1975 and 2030 occurs in the developing Regions IV (LA), V (Af/SEA), VI (ME/NAf), and VII (C/CPA). The population assumptions for both scenarios are identical.

1. For a description of the regions see Appendix, p. 165.

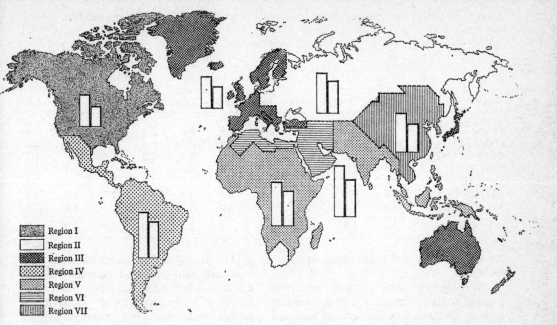

Region I
Region II
Region III
Region IV
Region V
Region VI
Region VII

*Economic growth.* Figure 1 shows the average 1975–2030 gross domestic product (GDP) growth rates assumed for each of the seven different regions for the two scenarios. These averages, however, mask an important characteristic of both scenarios—that in all regions of the world, the rate of economic growth continually decreases. The more detailed data are given in Table 2 along with historic growth rates for the periods 1950–60 and 1960–75. Except for the case of Region II (SU/EE) in comparison with Region VII (C/CPA), the growth rates in the developing regions consistently exceed those in the developed regions, although never by much. That the gap is not larger reflects a recognition that, for the next several decades at least, the devel-

Fig. 1. The assumptions about average growth rates of GDP for the IIASA high (left bar in each region) and low (right bar in each region) scenarios. The figures shown indicate percentage growth per year averaged for the period 1975–2030 for each of the seven regions.

oping countries will still be tied to the economies of the rest of the world through trade and other institutional relations.

Unlike the population assumptions presented earlier, the economic growth rates of Table 2 do not represent initial assumptions that remained unchanged throughout the subsequent analysis. They are rather the result of several revisions designed to ensure their consistency with the evolution of energy de-

147

TABLE 2. Historical and projected growth rates of GDP for the IIASA high and low scenarios (per cent/year)

| | Historical | | Scenario projection | | | | | | | |
| | | | 1975–85 | | 1985–2000 | | 2000–15 | | 2015–30 | |
| Region | 1950–60 | 1960–75 | High | Low | High | Low | High | Low | High | Low |
|---|---|---|---|---|---|---|---|---|---|---|
| I (NA) | 3.3 | 3.4 | 4.3 | 3.1 | 3.3 | 2.0 | 2.4 | 1.1 | 2.0 | 1.0 |
| II (SU/EE) | 10.4 | 6.5 | 5.0 | 4.5 | 4.0 | 3.5 | 3.5 | 2.5 | 3.5 | 2.0 |
| III (WE/JANZ) | 5.0 | 5.2 | 4.3 | 3.2 | 3.4 | 2.1 | 2.5 | 1.5 | 2.0 | 1.2 |
| IV (LA) | 5.0 | 6.1 | 6.2 | 4.7 | 4.9 | 3.6 | 3.7 | 3.0 | 3.3 | 3.0 |
| V (Af/SEA | 3.9 | 5.5 | 5.8 | 4.8 | 4.8 | 3.6 | 3.8 | 2.8 | 3.4 | 2.4 |
| VI (ME/NAf) | 7.0 | 9.8 | 7.2 | 5.6 | 5.9 | 4.6 | 4.2 | 2.7 | 3.8 | 2.1 |
| VII (C/CPA) | 8.0 | 6.1 | 5.0 | 3.3 | 4.0 | 3.0 | 3.5 | 2.5 | 3.0 | 2.0 |
| World | 5.0 | 5.0 | 4.7 | 3.6 | 3.8 | 2.7 | 3.0 | 1.9 | 2.7 | 1.7 |

mand and supply that is calculated to follow from them.

*Technological progress and structural changes within economies.* For these two categories, which include the sorts of technical and social changes usually labelled conservation, it is more difficult to summarize all the scenario assumptions in a few graphs or tables. As an indication of the extent to which energy conservation assumptions are reflected in the two scenarios, Figures 2 and 3 therefore present some of the aggregate results of the scenarios.

Figure 4 is a schematic representation of the IIASA set of energy models as they were used in constructing the scenarios.

The analysis began with assumptions belonging to each of the four categories just mentioned: population growth, economic growth, technological progress, and structural changes within economies. An Energy Demand Model then calculated for each of the seven regions the resultant evolution of *final* energy demand from 1980 to 2030.

The projected final energy demands were translated into projected secondary energy demands, which were then input to an energy supply and conversion model. Other inputs to this model were, first, assumptions constraining energy supply and conversion possibilities (see Figure 4) and, second, the results of a procedure analysing the patterns and prices of oil imports and exports among the seven regions.

The energy supply and conversion model calculated the primary fuel supplies and conversion facilities needed to meet the projected secondary energy demands at lowest cost and within the specified constraints.

Associated with providing the resources and facilities indicated by the energy supply and conversion model, there are necessarily direct and indirect requirements for capital, materials, manpower, equipment, land, water, and additional energy. In particular, capacities within crucial mining and manufacturing industries have to be expanded. These related

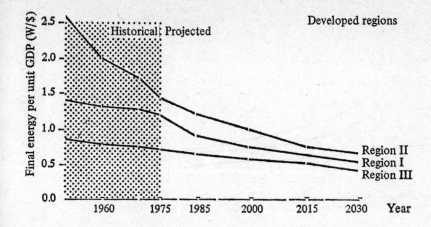

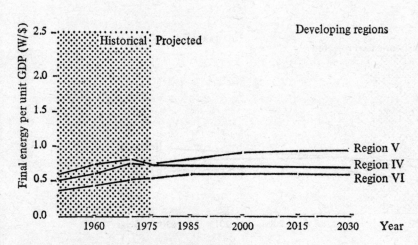

FIG. 2. Final energy per unit of GDP for the
high scenario in developed and developing
regions.

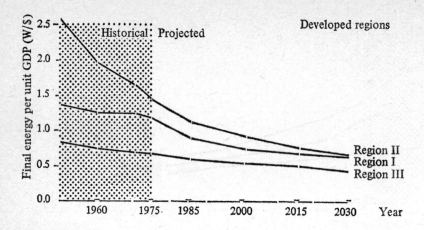

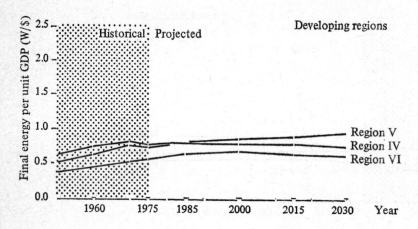

FIG. 3. Final energy per unit of GDP for the
low scenario in developed and developing
regions.

FIG. 4. A simplified representation of the IIASA
set of energy models used in constructing the
scenarios.

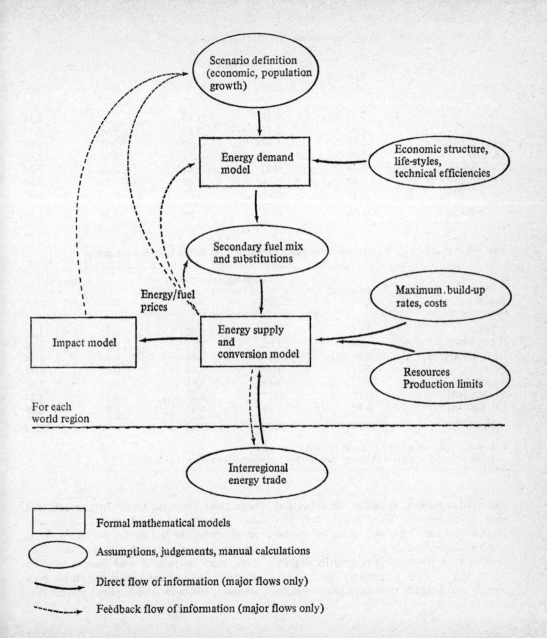

Scenario definition
(economic, population
growth)

Energy demand
model

Economic structure,
life-styles,
technical efficiencies

Secondary fuel mix
and substitutions

Maximum build-up
rates, costs

Energy/fuel
prices

Impact model

Energy supply
and
conversion model

Resources
Production limits

For each
world region

Interregional
energy trade

Formal mathematical models

Assumptions, judgements, manual calculations

Direct flow of information (major flows only)

Feedback flow of information (major flows only)

TABLE 3. Per capita final energy consumption (kWyr/yr) calculated from the scenarios

| Region | Base year 1975 | High scenario | | Low scenario | |
|---|---|---|---|---|---|
| | | 2000 | 2030 | 2000 | 2030 |
| I (NA) | 7.89 | 9.25 | 11.63 | 7.95 | 8.37 |
| II (SU/EE) | 3.52 | 5.47 | 8.57 | 4.98 | 6.15 |
| III (WE/JANZ) | 2.84 | 4.46 | 5.70 | 3.52 | 3.90 |
| IV (LA) | 0.80 | 1.75 | 3.31 | 1.28 | 2.08 |
| V (Af/SEA) | 0.18 | 0.42 | 0.89 | 0.32 | 0.53 |
| VI (ME/NAf) | 0.80 | 2.34 | 4.64 | 1.76 | 2.46 |
| VII (C/CPA) | 0.43 | 0.93 | 1.87 | 0.64 | 0.93 |
| World | 1.46 | 1.96 | 2.86 | 1.58 | 1.83 |

TABLE 4. Primary energy requirements by region (TWyr/yr) calculated from the scenarios

| Region | Base year 1975 | High scenario | | Low scenario | |
|---|---|---|---|---|---|
| | | 2000 | 2030 | 2000 | 2030 |
| I (NA) | 2.65 | 3.89 | 6.02 | 3.31 | 4.37 |
| II (SU/EE) | 1.84 | 3.69 | 7.33 | 3.31 | 5.00 |
| III (WE/JANZ) | 2.26 | 4.29 | 7.14 | 3.39 | 4.54 |
| IV (LA) | 0.34 | 1.34 | 3.68 | 0.97 | 2.31 |
| V (Af/SEA) | 0.33 | 1.43 | 4.65 | 1.07 | 2.66 |
| VI (ME/NAf) | 0.13 | 0.77 | 2.38 | 0.56 | 1.23 |
| VII (C/CPA) | 0.46 | 1.44 | 4.45 | 0.98 | 2.29 |
| Total[1] | 8.21[2] | 16.84 | 35.65 | 13.59 | 22.39 |

1. Columns may not add up to totals because of rounding.
2. Includes 0.21 TWyr/yr of bunkers—fuel used in international shipments of fuel.

industrial capacities, as well as the direct and indirect requirements listed above, were calculated for each scenario using a model labelled simply 'impact'.

This short summary is necessarily slightly misleading in that it presents the models linearly and suggests that the analysis simply began with the input for the first model, used each in turn, and ended up with some final results from the last model. In reality, as is usually the case with such sets of models, they were used in parallel and iteratively. The objective was internal consistency within each scenario, which in turn required several iter-

TABLE 5. Final energy growth rates for 1950–75 and projections to 2030 (per cent/year) calculated from the scenarios

| Region | Historical 1950–75 | High scenario | | Low scenario | |
|---|---|---|---|---|---|
| | | 1975–2000 | 2000–2030 | 1975–2000 | 2000–2030 |
| I (NA) | 2.7 | 1.4 | 1.1 | 0.8 | 0.5 |
| II (SU/EE) | 5.2 | 2.5 | 1.8 | 2.2 | 1.0 |
| III (WE/JANZ) | 4.3 | 2.6 | 1.2 | 1.7 | 0.7 |
| IV (LA) | 6.8 | 5.6 | 3.3 | 4.3 | 2.8 |
| V (Af/SEA) | 6.7 | 5.9 | 3.7 | 4.7 | 2.9 |
| VI (ME/NAf) | 10.4 | 7.0 | 3.5 | 5.8 | 2.3 |
| VII (C/CPA) | 10.8 | 4.7 | 3.2 | 3.1 | 2.1 |
| World | 4.3 | 3.0 | 2.2 | 2.1 | 1.4 |

ations of the model set. The major consistency checks between models are suggested by the dotted lines in Figure 4.

The resultant final per capita energy consumption in each region of the world is shown in Table 3 for both the high and low scenarios. Table 4 shows the corresponding primary energy requirements. Whether expressed in terms of final energy consumption or primary energy requirements, the results of both scenarios indicate a noticeable reduction in the gap between the energy budgets of the developed regions and those of the developing regions. The reduction is greater in the high scenario, where, because of overall higher economic growth rates, the developing regions are able to catch up more than they do in the low scenario. Still, in both scenarios, the advances achieved in 2000 and 2030 by the developing countries lie well below their currently expressed aspirations. For example, even in the high scenario, the 2030 per capita final energy consumption in Region IV (LA) remains below the 1975 level in Region II (SU/EE). And Region V (Af/SEA) has by 2030 only just passed where Region IV (LA) was in 1975.

As shown in Table 5, energy consumption growth rates decrease throughout the scenarios for all regions, though once again there is a noticeable difference between developed and developing regions. Part of this is due to the lower economic growth rates assumed for the developed regions, but part of it is simply because regions that use more energy today have more opportunities to conserve.

The resultant contributions of each primary energy source towards meeting the projected demand levels are shown in Table 6. For both scenarios the level of use increases for each source of primary energy. Most importantly, this includes the fossil sources—coal, gas and especially oil. For, although the share of primary energy requirements that is met by oil decreases in both scenarios (Fig. 5), the absolute amount of oil used increases.

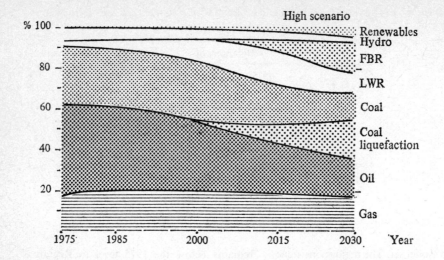

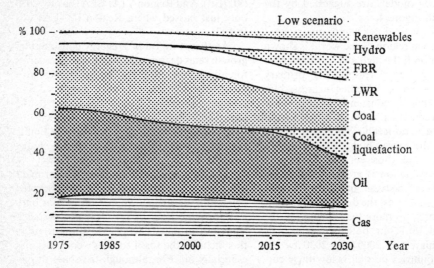

FIG. 5. The global primary energy shares by
source for the high and low scenarios.

| Primary source | Base year 1975 | High scenario | | Low scenario | |
|---|---|---|---|---|---|
| | | 2000 | 2030 | 2000 | 2030 |
| Oil | 3.83 | 5.89 | 6.83 | 4.75 | 5.02 |
| Gas | 1.51 | 3.11 | 5.97 | 2.53 | 3.47 |
| Coal | 2.26 | 4.94 | 11.98 | 3.92 | 6.45 |
| Light-water reactor | 0.12 | 1.70 | 3.21 | 1.27 | 1.89 |
| Fast-breeder reactor | — | 0.04 | 4.88 | 0.02 | 3.28 |
| Hydroelectricity | 0.50 | 0.83 | 1.46 | 0.83 | 1.46 |
| Solar | — | 0.10 | 0.49 | 0.09 | 0.30 |
| Other[1] | — | 0.22 | 0.81 | 0.17 | 0.52 |
| TOTAL[2] | 8.21 | 16.84 | 35.65 | 13.59 | 22.39 |

1. Includes biogas, geothermal and commercial wood use.
2. Columns may not add up to totals because of rounding.

However, the oil used in 2030 is very different from that used today. Figure 6 shows how the scenarios project that the future primary liquids demand of the regions with non-centrally planned economies will be met. Except for oil from Region VI (ME/NAf), none of the oil used after 2010 comes from currently known reserves of conventional oil. And by 2030 the portion of the primary liquids demand that is met by conventional oil reserves, including those yet to be discovered, is small. For the world as a whole, Figure 7 describes essentially the same story, though using slightly different terms.

Even with the projected increases in oil production of all sorts, Figures 6 and 7 indicate that in the twenty-first century the scenarios project an increasing gap between the demand for liquid fuels and the supply of oil. The gap is filled by liquefying coal at a rapidly increasing rate, as shown in Figure 8.

Two questions are immediately raised by these results. Why do the fossil fuels continue to dominate the world's energy system so persistently? And, given this fact, how much fossil fuel is left in 2030, according to the scenarios?

To the first question, two partial answers can be offered. First, there is the steadily increasing demand for liquid fuels, although both scenarios assume that in the future they will increasingly be reserved for essential needs (such as transportation and chemical feedstocks). In a sense, the demand for liquid fuels constitutes the key problem within the energy problem. Second, the rates at which new technologies can replace older, more inefficient users of fossil fuels are limited. Figure 9, for example, indicates that even by 2030 coal used for generating electricity is far from having been replaced by its theoretically unlimited non-fossil competitors, nuclear and solar power.

The answer to the second question, 'How

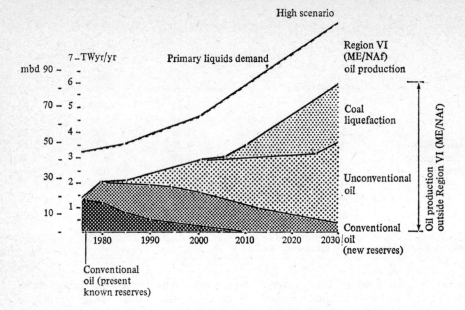

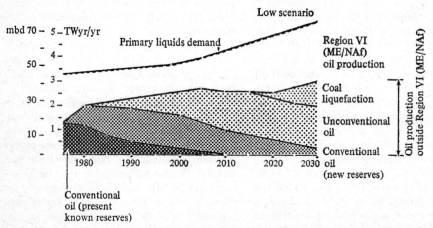

FIG. 6. The oil supply and demand calculated for
the regions with non-centrally planned economies
from the high and low scenarios.

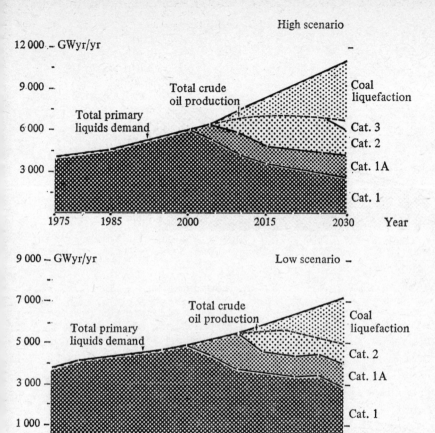

FIG. 7. The global oil supply and demand
calculated from the high and low scenarios. The
categories are for oil and gas: Category
1, $12/boe or less; Category 2, $12–20/boe;
Category 3, $20–25/boe. For coal: Category
1, $25/tce or less; Category 2, $25–50/tce.
(Category 1A includes oil at $12–16/boe.)

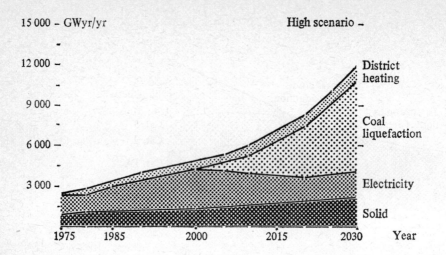

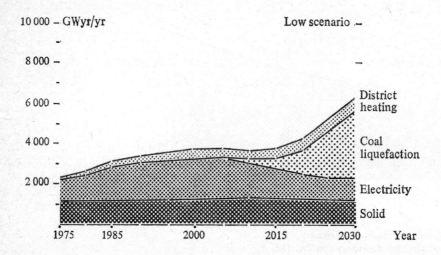

FIG. 8. The global coal supply and demand calculated from the high and low scenarios.

TABLE 7. The cumulative uses of fossil fuels

| Resource | Total resource available (TWyr) | Total consumed by 2030 as percentage of total available | |
| --- | --- | --- | --- |
| | | High scenario | Low scenario |
| *Oil* | | | |
| Categories 1 and 2 | 464 | 68 | 57 |
| Category 3 | 373 | 1 | — |
| *Gas* | | | |
| Categories 1 and 2 | 408 | 49 | 36 |
| Category 3 | 130 | — | — |
| *Coal* | | | |
| Category 1 | 560 | 61 | 40 |
| Category 2 | 1 019 | — | — |

much fossil fuel is left in 2030?' is given in Table 7. There is, according to the scenarios, quite a bit left, but it is not cheap, either financially, environmentally or socially. And at the ever-increasing consumption rates of the scenarios—already at 22.4 TWyr/yr to 35.7 TWyr/yr in 2030—it will not last for ever. Again the scenarios' message is the same. During the next fifty years the crucial constraint is not likely to be the availability of resources; rather it will be time: the time needed to reduce the demand for liquid fuels and the time it takes non-fossil technologies to penetrate the primary energy market.

Two important economic interpretations of the scenario results are displayed in Table 8 and Figure 9. The table shows how the projected growth in final energy consumption compares with the economic growthrates that were assumed at the beginning of the scenarios. The comparison is in terms of the final energy to gross domestic product elasticity. The higher this number, the faster final energy use is growing in relation to the economy as a whole. If the value is greater than 1, final energy use is growing faster than the economy; if the value is less than 1, the economy is growing faster. The numbers show that, as a general rule, as the scenarios move from 1975 to 2030, less and less energy is needed to fuel economic growth; that is, the societies of the scenarios are becoming ever more conservationist. The only exception to this trend is Region I (NA), because of its currently tremendous potential for conservation. The scenarios assume that this potential will be exploited quickly; in fact, a large part of the conservation occurring before 2000 is due simply to already mandated improvements in the fuel efficiencies of Region I (NA) automobiles.

The second important message of Table 8 is that, the more developed an economy is, the

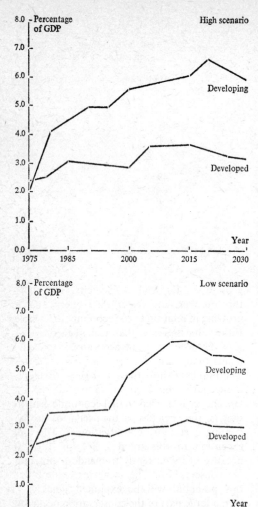

FIG. 9. Direct plus indirect energy investments as a share of GDP for developed and developing regions.

less energy it requires for economic growth. The elasticities for the developed Regions I (NA), II (SU/EE), and III (WE/JANZ) are all below 1 (the economy grows faster than final energy use), while for the developing regions the elasticities are predominantly greater than 1 (final energy use grows faster than the economy).

The capital investments required to support the expanding energy supplies of the scenarios are indicated in Figure 9, which shows the percentages of the gross domestic product that must be invested in energy facilities. As might be expected, the situation is most difficult in the developing countries, where, in the high scenario for example, energy investments peak at 6.6 per cent of GDP around 2020.

In addition to the two benchmark scenarios, three alternative cases were also analysed, though in less detail. As suggested by their titles, two of these—the nuclear moratorium case and the enhanced nuclear case—involved major changes in the assumptions concerning energy supply. The third alternative arose from a major change on the demand side: What are the implications if we assume that the global primary energy demand in 2030 does not exceed 16 TWyr/yr (compared with the low scenario value of 22.4 TWyr/yr)? The key characteristics of the three alternative cases turned out to be as follows.

The nuclear moratorium case indicates that the energy demands of the low scenario can indeed be met without new nuclear capacity. Fossil-fuel supplies are depleted more alarmingly than in the low scenario, gas assumes an especially important role, and solar electric power expands at its maximum rate. Needs are met, but costs are higher than in the low scenario (Fig.

TABLE 8. Final energy to GDP elasticities, 1950–2030, for the IIASA high and low scenarios

| Region | His-torical 1950–75 | Scenario projection | | | | | | | |
|---|---|---|---|---|---|---|---|---|---|
| | | 1975–85 | | 1985–2000 | | 2000–15 | | 2015–30 | |
| | | High | Low | High | Low | High | Low | High | Low |
| I (NA) | 0.84 | 0.31 | 0.24 | 0.43 | 0.38 | 0.53 | 0.53 | 0.48 | 0.46 |
| II (SU/EE) | 0.68 | 0.59 | 0.54 | 0.58 | 0.57 | 0.52 | 0.50 | 0.53 | 0.41 |
| III (WE/JANZ) | 0.84 | 0.77 | 0.67 | 0.65 | 0.64 | 0.58 | 0.60 | 0.51 | 0.49 |
| IV (LA) | 1.21 | 1.07 | 1.10 | 1.01 | 1.03 | 0.97 | 0.95 | 0.90 | 0.88 |
| V (Af/SEA) | 1.42 | 1.20 | 1.19 | 1.08 | 1.12 | 1.05 | 1.14 | 1.01 | 1.06 |
| VI (ME/NAf) | 1.17 | 1.12 | 1.21 | 1.07 | 1.11 | 0.95 | 1.01 | 0.81 | 0.93 |
| VII (C/CPA) | 1.53 | 1.10 | 1.02 | 1.02 | 0.98 | 1.02 | 0.99 | 0.96 | 0.90 |
| World (net) | 0.87 | 0.69 | 0.64 | 0.73 | 0.73 | 0.78 | 0.79 | 0.77 | 0.74 |

10). As the cushion of fossil fuels is diminished, time therefore becomes a tighter constraint.

In the enhanced nuclear case, in which energy demand is assumed to be at the same levels as in the high scenario, the unsettling depletion of fossil resources, so characteristic of both scenarios and the nuclear moratorium case, is abated only slightly. Despite the use of nuclear-generated hydrogen to produce liquid fuels from coal efficiently, in 2030 only 14 per cent of the liquid fuels produced are of nuclear origin, and the overall share of the global primary energy market held by nuclear power is only 29 per cent. Although this is higher than its 23 per cent market share in the high scenario, it remains below the 2030 potential of 40 per cent. The required investments are slightly higher than those of the high scenario (Fig. 11).

The case in which primary energy demand reaches only 16 TWyr/yr in 2030 necessarily implies zero growth during the next fifty years in the world's average per capita energy consumption. This is because 16 TWyr/yr represents a doubling in the world's primary energy consumption over the 1975 to 2030 period, which corresponds precisely with the projected doubling of the world's population during the same period: from 4,000 million in 1975 to 8,000 million in 2030. For the developing countries this case assumes, however, that projected energy uses are essentially as in the low scenario. The result is that per capita energy consumption particularly in Regions I (NA) and III (WE/JANZ) must drop significantly. Mathematically this can be accomplished by imposing a variety of assumptions: faster technological advances, a more rapid economic shift away from heavy industries and towards the service sector, reducing the projected uses of cars and planes, no air-conditioning at all in Region III (WE/JANZ), and so forth.

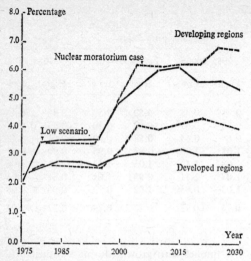

Fig. 10. The total energy investment as a share of GDP for the low scenario and the nuclear moratorium case.

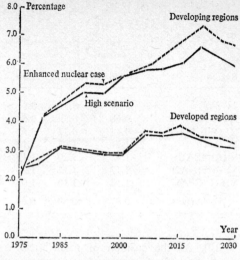

Fig. 11. The total energy investment as a share of GDP for the high scenario and the enhanced nuclear case.

Whether these mathematical manipulations could be reproduced in the real world by manipulating taxes, regulations, prices, subsidies, and all the rest is an open question. It is clear, however, that the carefully co-ordinated effort required would be unprecedented.

Hidden within all the aggregate results presented so far are crucial differences among the regions. Of these the most important are as follows.

*Region I (NA)*. Here the future of the scenarios is dominated by three considerations: a post-industrial, mature-economy slow-down; substantial energy savings because of technological advances and some restructuring of economic activities; and a rapid build-up of a coal-liquefaction industry to replace domestic and imported oil. None of these changes, except possibly the last, would be expected to produce profound or sweeping changes in the life-styles of North Americans.

The conservation effort envisioned includes, in particular, automobiles averaging 100 km per 7 l in 2030, homes 40 per cent more efficient in terms of heat loss than in 1975, and solar collectors attached to 50 per cent of all post-1975 single-family dwellings.

On the supply side, by 2030 Region I (NA) is neither a net importer nor a net exporter of oil. In the low scenario it is also self-sufficient in coal, and in the high scenario coal exports in 2030 equal 750 GWyr/yr.

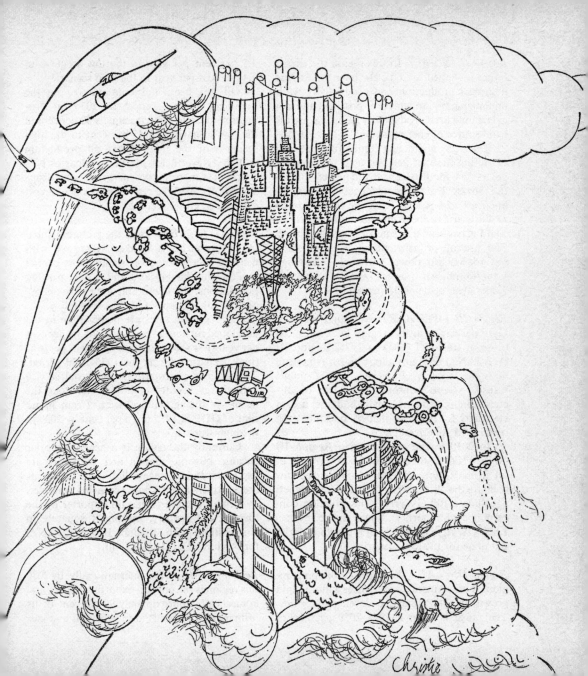

*Region II (SU/EE).* In this region the energy future is shaped by the clear intent to expand industrial production and productivity while minimizing oil use wherever possible, and it is in fact industrial productivity gains that are the main source of energy savings.

Through minimizing liquid fuel use and exploiting the vast gas and coal resources of Soviet Asia for district heat and power supplies the Soviet Union avoids becoming an oil importer. Oil exports from the Soviet Union to Eastern Europe continue, and exports of coal and gas from the region as a whole expand. The use of primary energy sources shifts towards nuclear power and coal. In the high scenario nuclear power's share in 2030 is 33 per cent; coal's is 38 per cent.

*Region III (WE/JANZ).* Although levels of GDP per capita in this region in 2030 exceed those of Region I (NA) in 1975, Region III (WE/JANZ) does not adopt North American life-styles entirely. Extensive use of public transit systems continues, the use of air-conditioning remains small, and the use of electricity for home appliances does not even reach 1975 American levels.

In 2030 in the high-scenario Region III (WE/JANZ) is still importing 600 GWyr/yr of oil from Region VI (ME/NAf). In the low scenario the oil imports in 2030 are even higher, equalling 1,100 GWyr/yr. The situation concerning coal imports is, however, the reverse: 1,600 GWyr/yr for the high scenario in 2030, but none for the low scenario.

*Region IV (LA).* Like other developing regions, Latin America experiences a more rapid growth in GDP than the developed regions. The range is from a 1975–2030 average of 3.5 per cent per year in the low scenario to 4.4 per cent per year in the high scenario.

Oil has been and continues to be the dominant energy source. In 2030 oil production in Latin America equals 30 per cent in the low scenario and 45 per cent in the high scenario of the total global oil production in 1976. Nevertheless, in both scenarios oil's share of the primary energy supply drops slightly. By 2030 the region is no longer an oil exporter in either scenario.

*Region V (Af/SEA).* Here the picture painted by the scenarios is bleakest. Endowed with neither energy-resource riches nor capital wealth, while having large and rapidly growing populations, the favourable long-term energy options for Region V (Af/SEA) seem few.

GDP growth rates are higher than in the developed regions, averaging from 3.3 per cent per year (low scenario) to 4.3 per cent per year (high scenario) during the 1975–2030 period. The current shift towards the industrial, service and energy sectors continues, as does the decline of the agricultural sector (from 36 per cent of GDP in 1975 to 16 per cent by 2030 in the high scenario).

Currently the region is a net oil exporter because Nigeria, Gabon and Indonesia are exporters and aggregate liquid-fuel demands are relatively low. In both scenarios, however, the region becomes a net oil importer by the turn of the century, putting it squarely in competition with Region III (WE/JANZ) for Region VI's (ME/NAf) oil.

*Region VI (ME/NAf).* Internationally, by 2030 this region is the only oil-exporting region, and domestically approximately 90 per cent of its primary energy needs are met by oil and gas.

Its economic growth rates are the highest of the developing regions at 3.6 per cent per year (low scenario) and 5.1 per cent per year (high scenario) as 1975–2030 averages. In the high scenario this leads to GDP per capita levels in 2030 that exceed those of Region I (NA) in 1975.

*Region VII (C/CPA)*. GDP growth rates in this region are high, but so are population growth rates. By 2030, GDP per capita levels reach approximately those of Region IV (LA) in 1975. The region remains neither an importer nor an exporter of energy. Its domestic oil supplies are effectively exhausted around 2030 in both scenarios, thus requiring increasing coal production and coal liquefaction in the twenty-first century. In the high scenario, coal production in 2030 reaches 3.2 TWyr/yr, as compared with 0.45 TWyr/yr in 1975.

This summary of scenario results is not intended either as a prediction of the future or a prescription for solving the world's energy problem. Rather it reports an exercise designed to provide insights and a better understanding of the long-term global nature of that problem. The objective was simply to detail the engineering and economic consequences that might follow from several different sets of reasonable assumptions. Thus the futures described in the two scenarios and three alternative cases do not chart a path towards any special goal; more particularly, they do not chart a path towards a sustainable global energy system. That is the concern of Chapter 15 which follows.

# Appendix: Regions used in the scenarios

Region I—North America (NA), developed market economies, rich in resources.

Region II—Soviet Union and Eastern Europe (SU/EE): developed, centrally planned economies, rich in resources.

Region III—Western Europe, Australia, Israel, Japan, New Zealand, and South Africa (WE/JANZ): developed market economies, poorer in resources than the other developed regions.

Region IV—Latin America (LA): a developing region with market economies and many resources.

Region V—South and South-East Asia, and sub-Saharan Africa excluding South Africa (Af/SEA): developing regions with mostly market economies, but with relatively few resources (except for some notable exceptions, e.g. Nigeria and Indonesia).

Region VI—Middle East and North Africa (ME/NAf): a special case with their economies in transition and with rich oil and gas resources.

Region VII—China and other Asian countries with centrally planned economies (C/CPA): developing regions with only modest resources.

# Paths to a sustainable future

## 15

Alan McDonald

*An analysis that takes energy demand into account soon reveals disadvantages in the nuclear- and solar-power options. The best alternative may be some chemical energy carrier—hydrogen, for example. But whatever the ultimate world energy system, it will be reached only along a path of continued expansion of productive capabilities and continual learning how to use limited resources more ingeniously.*

There is an underlying unifying theme running throughout the work that, while not surprising in hindsight, was not obvious at the beginning. It has to do with the general pattern of the world's response to the increasing scarcity and expense of energy resources.

As we have become more aware of the problems of energy resources throughout the 1970s, we have begun to adapt in ways that make better use of the limited energy currently available. Sometimes we label these adaptations conservation; sometimes we call them improvements in efficiency; sometimes they are referred to as productivity increases. Whatever we call them, they all involve reducing the energy needed to produce some service (be it a well-heated sitting room or intercity jet travel) by replacing it with something else. In some cases this replacement is in the form of capital resources (e.g. investing in home insulation); in others it can be classified as labour

(e.g. periodic tuneups of an automobile to increase its fuel mileage); and in still others it may be labelled simply ingenuity or know-how (e.g. anything from more carefully planned shopping trips to large-scale reconfigurations of industrial processes).

At a personal level, we are all familiar with such adaptations—such substitutions of capital, labour, or know-how for energy in producing services. At more collective levels, ranging from small business enterprises to international alliances, we are becoming more familiar with them. And what will appear in the discussion to follow is the conviction that what may now seem to us to be perhaps quite sophisticated, energy-conserving arrangements of our resources of capital, labour, know-how, and energy indicate only the direction in which we can travel. They in no sense even begin to suggest the limits of what can be done.

Of particular importance is the notion of investing these resources to increase the stock available in each category. Again, these ideas are hardly unfamiliar—investments in education, in research and development, in capital equipment,

An edited version of Section 5: 'Paths to a Sustainable Future' (IIASA, 1981).

in exploratory drilling have all contributed, and continue to contribute, to the resources that we can put to use. What is less familiar is what these same concepts lead us to when applied from a global perspective contemplating the next half century and beyond.

Nuclear fission, nuclear fusion, and hard solar power were described in Chapter 2 as possible bases for a sustainable energy system. However, that analysis ignored entirely the question of energy demand—in what forms energy will in fact be needed—and from this perspective it is clear that nuclear and solar power are not without their disadvantages. As generally conceived, they produce energy in the form of heat that is assumed to be used directly sometimes or, more often, converted to electricity. And both heat and, to a lesser extent, electricity have their drawbacks: they are difficult to store and to transport. It is for these reasons that, in situations where favourable storage and transport characteristics are particularly important, we have tended to rely on chemical energy carriers, principally in the form of fossil fuels.

It is precisely these fossil energy carriers, however, that are getting scarcer. While electricity can replace them to some degree, for the reasons listed above we might be better off developing an alternative that is itself a chemical, rather than an electrical, energy carrier.

A possible candidate is hydrogen. It is attractive, first, because the technology for converting electricity to hydrogen via the electrolysis of water is well developed. Second, processes for converting nuclear or solar heat directly to hydrogen without the intermediate step of electricity production appear promising. Third, hydrogen is much more easily stored than electricity and might be particularly suited to large-scale storage in depleted natural gas reservoirs. Fourth, the piping networks and the infrastructure associated with further large-scale use of natural gas would be especially suited to a gradual replacement of natural gas by hydrogen. And fifth, when hydrogen is burned (recombined with oxygen), it produces essentially only water vapour, thus making its use environmentally attractive.

To introduce hydrogen on a scale comparable to that of electricity cannot be done overnight; but it can most certainly be started in a way that contributes from the beginning to solving the critical, immediately pressing liquid-fuel problem. Consider, first, hydrogen production. Among the many possible production processes there are some that begin by converting methane (the principal component of natural gas) into methanol, a liquid hydrocarbon fuel. Thus, even to introduce just the first step of such processes would be to introduce a capability to convert natural gas to liquid fuel, to convert a large resource that is often wasted (flared) because of the world's currently undersized long-distance gas transportation system into a form in which it can be much more easily transported, stored and used.

Once hydrogen is available, moreover, it can be used to exploit more efficiently the most extensive of the fossil fuels, coal. Specifically it would allow the introduction of allothermal coal-liquefaction schemes, as discussed in Chapter 2. Thus, hydrogen produced by nuclear or solar facilities, in conjunction with the heat generated by the same facilities, could be used to extend by a factor of three to four the portion of our coal resources that could be devoted to producing liquid fuels. In view

Christie

of the results of Chapter 14, such an extension could be critical.

Still, to produce and use synthetic liquid hydrocarbons from coal and other fossil resources is to consume the store of carbon atoms available in these particularly convenient forms. If liquid hydrocarbons are therefore to play any kind of role in a sustainable energy system, the problem that must be solved is that of recycling carbon, of extracting carbon dioxide from the atmosphere and combining it, rather than fossil fuels, with hydrogen to produce liquid hydrocarbons. The simplest way to exploit the carbon reservoir in the atmosphere is to use plants which are already extracting carbon dioxide continuously. Much technology for converting biomass into liquid fuels has been developed, and here again, external sources of hydrogen and heat can help conserve the carbon resource. A more direct way to conserve the carbon atoms incorporated in synthetic hydrocarbons is to capture immediately the combustion gases released when the fuel is burned and then recycle them.

These are only suggestions. They are motivated by the effort to fill a crucial gap revealed by the analyses we have made, the gap between the immediate but apparently persistent demand for liquid hydrocarbons and the supply of heat and electricity that could be produced by the only possible sustainable energy sources: those based on nuclear and solar technologies. The central notion in filling this gap is the importance of using the world's store of carbon atoms prudently. The basic ideas are familiar: recycling and using coal as a bridge to the future. Only the scale is different: the continual recycling of the world's supply of carbon atoms forever, and the investment of our coal resources in building not only a bridge to the

twenty-first century, but quite possibly a bridge extending all the way to the twenty-second century.

Only the outlines of how the world's energy system might ultimately develop can be sketched here. How they might be filled in must be left for subsequent analysts, and whether the course suggested here will be pursued at all must be left for subsequent generations. It is clear, however, that building a sustainable energy system will require the continual expansion of the world's productive capabilities, in all dimensions. For orientation, even the low scenario of Chapter 14 would require between 1975 and 2030 an increase in the total capital stock of the world by a factor of approximately 20 to 30. This is why it is so important that the eight billion or so people living in 2030 should be rich, not poor, and much richer than today. That they should be rich does not mean that they must discover some new treasure of physical resources that has been completely overlooked in the study reported here; it means instead that they and their predecessors will have learned how to use the limited resources available more efficiently, more ingeniously, more productively. The process is continuous, and it is cumulative.

170

# A last word

## 16

Tom Mikkelsen

*A medical researcher examines the world's energy problem and makes some specific suggestions as to how we can alter our ways of doing things in order to prepare for a promising future.*

The safe and sufficient supply of energy is probably the most costly and complex of the issues to be resolved by *Homo cunctans*, procrastinating man, today on Spaceship Earth. How will we provide enough energy—without adverse effects on the human environment—beyond the year 2000? What kind of energy can keep our generators working beyond the twentieth and thirtieth centuries?

To most people in the industrialized world energy and petroleum are synonymous. Every day, 45 million barrels each containing 158.9 litres (or 42 United States gallons) of oil are transported from producers (mainly the thirteen OPEC members, and of these chiefly Saudi Arabia) to consumers, largely in the western hemisphere. Annually about 16,500 million barrels of oil are consumed from a total world reserve estimated in 1975 to be 658,000 million barrels.[1] Each year an additional 2 million tonnes of crude oil pollute the oceans and coastlines during transport,

loading and off-shore production.[2] If, to these 2 million tonnes, we add the amount of oil and related material ending in the sea that come from rivers, sewers, and the fall-out of hydrocarbons emitted on land by motor vehicles, the total amount of pollutants from such sources is 10 million tonnes annually.[3]

If the above-mentioned figures of demand and supply were frozen, another forty years would pass before the last drop of oil was used in the year 2018. The reserves are not all economically or technically recoverable, however, and undoubtedly demand will grow without the guarantee that supply can follow.

1. C. Wilson, *Energy: Global Prospects 1985–2000, Report of the Workshop on Alternative Energy Strategies*, p. 118, New York, McGraw-Hill, 1977.
2. W. Palz, *Solar Electricity, An Economic Approach to Solar Electricity*. London, Butterworth; Paris, Unesco, 1977. Also available in French, Italian and Spanish editions.
3. J. Main, *Pollutants: Poisons Around the World*, p. 9, The Conservation Foundation and Secretariat, United Nations Conference on the Human Environment, 1972. (Man's Home series.)

Originally published as 'Energy and the Options for Mankind', *Impact of Science on Society*, Vol. 29, No. 4, 1979, pp. 375–82.

It is likely that new oil fields will be discovered and exploited (as in Alaska and Mexico), and even though a political plug is not likely to be dropped in the OPEC well for the moment—although we have witnessed certain cutbacks in production for various reasons—it must be realized that the end of the era of growth in oil production is probably at the most only fifteen years ahead.[4] It should also be realized that society will have to find ways to manage the unavoidable transition from its present oil dependency to a far greater reliance on other energy sources.

## The problems with coal . . .

Coal, the 350-million-year-old deposits of naked seed (or gymnospermous) plants, could theoretically once again dominate the energy market. Economically recoverable coal has been estimated to amount to 737,000 million metric tonnes equivalent of approximately 3,000,000 million barrels of oil.[5] The ultimately recoverable world resources, according to the World Energy Conference, amount to 11,000,000 million tonnes (equivalent to 50,000,000 million barrels of oil). In other words, economically recoverable coal resources are at present calculated to be between four and five times the proven reserves of oil. At the present rate of energy consumption, 737,000 million tonnes of coal would last approximately 200 years. If necessary, coal could bridge the gap between today's oil and gas era and a future era of renewable energy resources. But we must not overlook the environmental impact that would follow a transition to coal, even for a few decades. We know little about the long-term impact on the atmosphere of fossil fuel combustion, but we do know that when coal is burnt the sulphur emission forms sulphur oxides which react with water to form acids which are dangerous to health and the environment. We know too little about clean-burning techniques—and too much about black-lung disease and other respiratory ailments among those who will have to work in the mines—to go ahead without careful, time-consuming research and planning. Problems regarding coal-sludge disposal will have to be solved in advance, since it has been calculated that the sludge generated in the United States over a twenty-year period would cover 500 km$^2$ to a depth of 1.5 m.[6] Although the cost of coal is at present relatively low, its extraction costs will probably increase fast, not least in Western Europe where mines are deeper and coal seams thinner than, for instance, in the United States. Finally, we should not forget that potential users, industrial as well as domestic, have been perfectly happy with 'clean', practical oil delivery for decades. It would require very strong motivation to convince consumers to adopt coal in place of oil, not least because coal is just another limited source of fuel with seriously adverse environmental effects.

Coal can be converted into petroleum (by the indirect, or Fischer-Tropsch liquefaction) or it can be gasified. But cost estimates of more than $16 a barrel for synthetic oil and more than $13 per barrel for methanol seem to be inhibitive combining with a lead-time of approximately ten years before a significant number of plants could be in operation.

4. Wilson, op. cit., p. 145.
5. Ibid., p. 171.
6. Palz, op. cit., p. 28.

# . . . and the perils of fission

I feel safe stating that there is no need for mankind to venture further into an era of nuclear fission power. Those in favour of nuclear fission plants (rightly) argue that time is running out for conventional fossil fuels; they (erroneously) insist that the fission technology is an unconditionally necessary development worldwide if we do not wish to witness a substantial part of our traditional materialistic civilization vanish in thin air like Shakespeare's witches.

At present three principal types of nuclear power reactors are in use: the light-water reactor (using either pressurized water or boiling water), the heavy-water reactor and the fast-breeder reactor. These reactor types carry out the same work as plants operating with fossil fuel. They generate steam which activates turbines attached to generators producing electricity. There is no magic about this.

In thermal reactors, the nuclear fuel consists of fissile nuclei of uranium ($U^{235}$). The chain reaction caused when the $U^{235}$ nucleus splits, following absorption of a neutron, is controlled (i.e. slowed down) by a moderator with light nuclei such as those of ordinary water or heavy water (deuterium oxide, $D_2O$).

The heavy-water reactor is dependent on natural uranium as fuel and heavy water as moderator. The uranium is used only once, and used fuel elements containing uranium, plutonium and radioactive wastes are stored—while we contemplate what to do with these substances in the future. At present, they are not reprocessed.[7]

For use in the light-water reactor, the uranium has to go through enrichment processes before fuel elements can be loaded in the reactors. The used-fuel elements are transported to reprocessing plants where they are treated mechanically and chemically to sort out (a) not only the remaining uranium and other radioactive fission products, but also (b), notably, the plutonium which for more than thirty years has been used to manufacture nuclear weapons. Plutonium may also be used as fuel-loading for the third reactor type, the fast-breeder reactor, which requires large amounts of plutonium initially. The fast-breeder operates with fast neutrons and 'breeds' more fuel than it consumes, converting non-fissile $U^{238}$ to fissionable plutonium by absorption of these fast neutrons.

There is a very real danger that plutonium-loaded nuclear devices could be used either by legitimate military forces or by militant terrorist groups. That it is indeed technically feasible for a talented amateur to build a nuclear bomb was proved by an American student some years ago and discussed throughout the world. Enough plutonium is already missing from 'safe' repositories of this element to manufacture several dangerous devices. As long as we have not met the general demand for equal opportunities for all and achieved peaceful settlement of conflicts, fear of terrorists will grow steadily as national and global stability diminishes—since no means for rendering plutonium useless for weapons is known.[8]

7. Wilson, op. cit., p. 196.
8. Ibid., p. 194.

## The difficulties with other energy sources

None of man's senses can detect radioactivity. Accordingly, we do not know how much damage has been inflicted already on mankind and his environment by the stocks of radioactive wastes scattered around the world as well as by numerous bomb tests. But we know enough of the potential dangers to work incessantly on the development of safe, renewable energy sources in order to render the nuclear fission plants obsolete before they spread further. An additional factor in favour of abandoning nuclear fission plants is that the world's deposits of high-grade uranium and thorium ore will probably be exhausted simultaneously with the exhaustion of petroleum.

Natural gas, as either associated gas (i.e. gas in solution with oil or in a gas cap above the oil) or unassociated gas (found in structures where only natural gas can be produced), is another substantial source of limited energy at present supplying 40 million households and 3 million businesses in the United States.

Total, proven world reserves of economically recoverable gas are equivalent to 386,000 million barrels of oil. The major share is controlled by the OPEC cartel. Moody and Geiger have estimated that the ultimately recoverable reserves are equivalent to 1,400,000 million barrels of oil.[9]

Our supply of natural gas is at present a minor problem compared to those of transport and distribution from well to consumer. The construction of pipeline networks is extremely costly. The alternative to pipelines, the transport of liquified gas by tanker, is not very attractive. Not only will enormous investments in liquefaction and regasification technology be required to carry the gas in special refrigerated tankers at —161 ºC, there is also a growing fear of the casualties that would follow the explosion of a tanker in port. There is reasonable doubt, as well, regarding the vast investments required for an energy source certain to run dry in the foreseeable future.

Shale oil extractable from rocks heated to about 500 ºC is not a realistic source of energy. Not only does it present refining problems (such as removing its nitrogen), but there is needed a tonne of rock to produce 100 litres of raw shale oil. Solid waste, estimated on the basis of a production rate of a million barrels per day, is calculated to be 1.7 million tonnes of rock daily.[10]

Geothermal energy in the form of hot brine has long been used in a number of countries (Iceland, Italy, Japan and the USSR) for domestic heating. This will undoubtedly continue to be an important source of energy locally.

## Water and plants as power sources

The indirect solar energies—water, wind and photosynthesis—have attracted renewed interest during the last decade. Hydroelectricity and wind energy are attractive for several reasons. They produce no chemical, thermal or radioactive pollution; they present

9. J. Moody and R. Geiger, 'Petroleum Resources: How Much and Where?', *Technology Review*, March 1975.
10. Palz, op. cit., p. 27.

no problems concerning fuel or transport; and they are renewable. Hydroelectric systems are well suited for the storage of electricity. It is therefore not surprising that countries with considerable mountainous area, such as Norway and Switzerland, obtain 100 per cent and 90 per cent respectively of their total electricity consumption from this source. In the United States and Western Europe, the greater part of the available potential has already been exploited, but vast areas of the developing world still have much potential power to extract from hydroelectricity.

Wind power generated by windmills has been exploited for hundreds of years, especially in the Netherlands where 20,000 windmills of 20 kW each were in operation at the end of the eighteenth century.

It has been estimated that about 50,000 windmills, having propeller diameters of 56 m and an average power of 500 W, would be required to produce energy equal to a million barrels of oil per day.[11] It is hard to imagine wind energy as a technology of energy supply. Local use will probably continue to be important in appropriate areas of the world, ideally combined with an electric storage system, either hydroelectric or electrochemical (batteries).

Various ways of extracting energy from plant matter have attracted interest. These are mainly based upon the fact that each year photosynthesis builds up and stores in plant matter seventeen times more energy than the world now consumes. So the growing of plants for fuel has been proposed.[12] Land-based energy plantations are potential competitors with agriculture and, considering the scarcity of cultivable land for a growing and increasingly hungry humanity, this exploitation

of land is not recommendable. Another type of bioconversion, namely the process of burning municipal waste or converting it into methane, is well known and highly useful in many parts of Europe; it could and should be extended to other parts of the world. The burning of residual straw and other agricultural wastes is increasingly common, not least in Denmark where more than 8,000 straw burners are in operation. These and other bioconversion technologies will continue to be important sources of energy locally. Necessary financial and educational support should be developed to expand these activities.

## The promises of solar energy and nuclear fusion

The limit to solar energy lies beyond the limits to life itself. This is an energy technology which can be supplied worldwide, without adverse effects on either micro- or macro-environment. Its technology is beyond major political or economic manipulation, thus making solar energy our most obvious choice for a safe and healthy future. We have already designed conversion systems for solar energy, for the direct conversion for heating of water and space domestically and industrially, as well as for (a) refrigeration and (b) production from heat of mechanical power (by means of an intermediate thermodynamic step) or electricity. The last process is usually called the photovoltaic effect.

In 1972, the world's total energy consumption was about $56 \times 10^{12}$ kWh, equal to the

11. Wilson, op. cit., p. 229.
12. Ibid., p. 228.

175

solar energy received every year by an area of 22,000 km² in a desert region. Such an area constitutes only about 0.005 per cent of the surface of the globe.[13] Including storage losses, non-productive spaces and losses in conversion efficiency, Wolfgang Palz has calculated that a solar energy system covering about a quarter of the area of Egypt could theoretically provide all the energy consumed in the world today.

Comparatively less is known about the future prospects of nuclear fusion power. (See the article on fusion (Chapter 4) in this volume.) I include it among the realistic future energy sources as an apparently attractive potentiality, largely because the elements of the basic fuel (deuterium and tritium, two heavy isotopes of hydrogen) are ubiquitous and virtually unlimited.

Fusion is the kind of nuclear reaction that converts mass into energy inside the sun, thus its technology is a matter of reproducing the sun's processes on earth. Since the oil embargo of 1973, research has been intensified substantially in this field. Significant progress has been accomplished, notably by Gerold Yonas and his colleagues at the Sandia Laboratories in Albuquerque[14] and by a group of scientists at the I. V. Kurchatov Institute of Atomic Energy in Moscow, among numerous others described in this book.

### Replicating the sun's processes

Two approaches towards nuclear fusion have been undertaken. These are: magnetic confinement schemes, where the gaseous fuel is held within the reactor chamber by means of strong magnetic fields; and inertial confinement, relying on powerful laser beams to implode fuel pellets of hydrogen isotopes. To replace the rather inefficient laser beams, researchers have undertaken a more promising approach to fusion through inertial confinement employing intense beams of electrons and ions generated by high-current, high-voltage electric pulses. These rather simple, high-energy electron-beam accelerators are produced at a cost of 2 per cent of the cost of lasers of comparable energy.

To produce steam for turbines with a heat source of $100 \times 10^6$ °C is a procedure at present only operational for milliseconds at a time. Apart from the tremendous problems presented (especially to metallurgists), intricate problems calling for the most advanced technology—and even technology not yet invented—must find their solutions before this energy source can be tamed for use. It is obvious that there must be a substantial risk of thermal pollution until we find ways to divert the enormous heat to industrial use. Since solar energy may not be sufficient or feasible to fulfil all human demand for energy, it is evident that the reproduction of the sun's processes on earth should be explored and financed intensively. This is so not only because fossil fuels, and uranium and thorium ores, will be exhausted in the foreseeable future, but also because the hazardous storage of radioactive wastes from fission reactors and the stockpiling of nuclear arms would be brought to an end.

As outlined by Yonas, nuclear fusion power is attractive for several reasons. A thimbleful of liquid heavy-hydrogen fuel would release

13. Palz, op. cit., p. 68.
14. G. Yonas, 'Fusion Power with Particle Beams', *Scientific American*, Vol. 239, No. 5, 1978, p. 40.

Christie

as much energy in the form of energetic neutrons as 20 tonnes of coal. Secondly, provided the materials for the walls of the reaction chamber are carefully selected, the fusion reactor would have fewer radioactive by-products than a fission reactor. Thirdly, there would be no possibility that the fuel ore would melt down, *in memoriam* Three-Mile Island.

## Man's apprehensiveness about innovation

Both solutions present innovation. Many people fear the challenge of innovation more than they fear the obvious shortcomings of the traditional energy strategies based on the exploitation of fossils or the plutonium-producing man-made radioactivities.

When Denis Papin built the first steam vessel in 1707, this was innovation. When James Watt perfected his own steam engine sixty years later, this was also innovation. When Hans Christian Oersted and Michael Faraday made the basic discoveries which eventually made possible modern electrical machines, society was curious but sceptical; and when the industrialist, Werner von Siemens, and Sir Charles Wheatstone some forty years later showed how to make a dynamo self-exciting, and Thomas Edison finally switched on municipal electric lighting by means of a power plant built in New York in 1882, the world was astonished. All these events were somewhat frightening innovations in the history of mankind.

Man constructed the internal combustion engine, making possible the automobile and the aeroplane. Man invented the turbine and subsequently the jet engine. He had discovered nuclear energy before Otto Hahn could tell us, in 1938, how to split the uranium atom

artificially, and before Enrico Fermi constructed the first nuclear reactor in Chicago in 1942. All these innovations were preceded by seemingly insurmountable challenges and problems. Eventually, all were surmounted. Some of the technical complexities concerning solar energy applied on the large scale necessary, and a number of those concerning the taming of nuclear fusion, seem to be insurmountable. But we shall be able to solve them through a global effort to do so.

We have seen the price of petroleum rise from $2.41 per barrel in 1973 to almost ten times this amount about seven years later. To me—as to most people, I suppose—it is not possible fully to comprehend the mechanisms which underlie pricing. World opinion tends to blame alternately the oil-producing countries, the 'seven sisters' of the oil industry, and the finite limits of oil for the soaring prices. Cutbacks in production are realities. Switching around supertankers while on the high seas in order to reach the port of maximum profit is a well-known tactic. The shortage of oil has encouraged individual nations to deal directly with the producing countries, thus short-circuiting the big multinational distributors.

# Back to the beginning

We must perceive all the elements making up the transition we are experiencing. We must make plans today for a realistic future, rather than hesitate until the future has suddenly become the present.

It is inevitable that burning of fossil fuel is related to serious adverse effects on our biosphere and thus on human health, especially in the congested and suffocating megalopolises. In spite of the SALT II agreement, nuclear-fission technology has supplied the so-called superpowers with nuclear weapons sufficient to destroy civilization several dozen times over. Insufficient security systems have closed down Harrisburg and other nuclear-fission plants, although we have not yet experienced a disastrous melt-down. Problems regarding 'safe' disposal of radioactive wastes have not been solved. How can they be solved, when we do not know where the next earthquake will occur?

Even the production of fossil fuel, i.e. excavating the crust and mantle of the planet, sooner or later must create problems regarding geological stability. In the United States, well over 500,000 oil wells are currently in operation. As fossil fuel runs out and nuclear-fission plants are renounced or closed for time-consuming security checks all over the industrialized world, we may be approaching the final discussion about our remaining fuel resources. Will this be a fierce and desolate discussion without words?

The birth of the universe and exact knowledge about its 'prenatal' and 'neonatal' events will most likely continue to be elusive subjects. The age of the universe has been estimated to be about 20,000 million years. The Milky Way, our galaxy, is possibly about 15,000 million years old; it has a diameter of 100,000 light-years and consists of 100,000 million stars.

Dating of certain types of stony meteorites by means of radioactive isotopes has made it possible to estimate the age of our planet as 4,600 million years. Aggregations of interstellar gases and dust particles—at varying distances from the core, or protosun, of the rotating primordial cloud—formed the larger, outer planets (Jupiter, Saturn, Uranus and Neptune) and the smaller, inner planets (Mercury, Venus, Earth and Mars). This probably took place some 5,000–6,000 million years ago, about the time when our sun began to shine.

Although our knowledge of the earliest history of the earth is at best fragmentary, we know that life was present approximately 3,400 million years ago. The earliest fossil traces of soft-bodied metazoans are burrows appearing in rocks younger than 700 million years. In other words, for 5,000–6,000 million years the earth has received solar radiation and metazoan life has flourished for 600 to 700 million years by means of this radiation. This life was also supported by a total biosphere exploited in a cautious and intelligent manner.

## Ways to solve the problem

Is *Homo scientificus* now going to disrupt his successful evolution? Will we let *Homo scientificus* destroy or seriously disable, in one frantic century, the results of millions of years of construction? Or are we going to act in accordance with the immense responsibility put on us all?

If we agree to honour a global mutualism and rationalism, we have excellent opportunities to house, feed and supply with energy the 6,200 million people we inescapably shall accommodate by the year 2000.

The dignified beginning of a new world order of such rational mutualism should be a unified effort to (a) share the fossil fuel we have left in an honest, minimum-profit manner in order to bridge the gap between today's oil-gas-coal era and a coming age of renewable energy supply; and (b) convert the necessary shares of the $400,000 million annually wasted on military absurdities and the technological and scientific brainpower devoted to military R&D to intensified research and development in the energy technologies suggested: primarily solar radiation and thermonuclear fusion.

It sounds very simple, and it is very simple. Merely a question of global common sense. The solution is so ridiculously simple that no one seems to have dared to propose it on a worldwide scale, the only valid scale now. I do not mind being ridiculed. Do you? I would like to invite well-documented, well-argued statements that, first, issues (a) and (b) cannot in fact be carried out, and secondly, that there are other solutions that will serve mankind significantly better in the long run—economically, ecologically, technologically, medically, politically and administratively.

I do not think we have time to invite obsolete arguments from ignorant politicians not possessing the necessary vision and courage, or from misguided scientists without a social conscience, or from individuals exclusively devoted to the promotion of personal or corporate profits. They have been heard too often, for too long a time.

We call ourselves Europeans, North Americans, Asians, Africans, etc. But we are more. We are human beings, a unique species on earth, where we have as our foremost, self-perpetuating duty to create decent lives for ourselves by creating it for others. We constitute one large species, *Homo sapiens, sapiens* prior to being black or white, European, Latin American or Australasian. We have long passed the day when ideological,

egocentric quarrel was relevant. We are in a new world—a small world—one in which worn-out national boots no longer fit to walk. Now we must succeed in creating the new man for this world. Otherwise, our once proud and vastly creative species will commit the most perfect and most deplorably futile suicide.

## Bibliography

BEHRMAN, D. *Science and Technology for Development, a Unesco Approach*. Paris, Unesco, 1979.

GEORGESCU-ROEGEN, N. *Energy and Economic Myths, Institutional and Analytical Economic Essays*. New York and Oxford, Pergamon Press, 1976.

MORAZÉ, C. *Science and the Factors of Inequality*. Paris, Unesco, 1979.

UNESCO. *An Introduction to Policy Analysis in Science and Technology*. Paris, 1979. (Science Policy Studies and Documents, 46.)

# Appendix: Energy units

There are two fundamental types of energy units, those that describe *amounts* of energy, and those that describe *rates* at which energy is supplied, converted, transported or used. In the first category, amounts, are units such as barrels of oil equivalent (boe), tonnes of coal equivalent (tce), or kilowatt-hours of electricity (kWh(e)). In the second category, rates, are million barrels of oil per day (mbd), tonnes of coal equivalent per year (tce/yr), and kilowatt-hours of electricity per year (kWh(e)/yr).

The most commonly used unit for amounts of energy is the terawatt-year (TWyr). One terawatt-year (1 TWyr) is equal to 1 million million watt-years (which can also be written as $10^{12}$ Wyr). It is therefore also equal to 1,000 million kilowatt-years ($10^9$ kWyr) or 1 million megawatt-years ($10^6$ MWyr) or 1,000 gigawatt-years ($10^3$ GWyr).

The most commonly used unit for rates of energy supply, conversion, transportation and use is the terawatt-year per year (TWyr/yr). The unit, terawatt (TW), which is sometimes used in place of terawatt-year per year (TWyr/yr), is reserved for the description of the capacities of various energy-conversion fa-

cilities. Thus the capacity of an electricity-generating station might be listed as 1,000 MW(e) (= 0.001 TW(e)). Since energy-conversion facilities seldom operate at their installed capacity all year long, their ratings in TW or GW or MW will differ from the actual rate at which they convert energy, as expressed in TWyr/yr or GWyr/yr or MWyr/yr.

Some particularly useful conversion factors are

1 TWyr = 30 quads ($30 \times 10^{15}$ British thermal units (BTU))

1 TWyr = 30 trillion cubic feet of gas ($30 \times 10^{12}$ ft$^3$ of gas)

1 TWyr = 1.1 thousand million tonnes of coal equivalent ($1.1 \times 10^9$ tce)

1 TWyr = 5.2 thousand million barrels of oil equivalent ($5.2 \times 10^9$ boe)

## Prefixes

tera (T)  = $10^{12}$
giga (G)  = $10^9$
mega (M) = $10^6$
kilo (k)  = $10^3$

Adapted from the surveys made by the International Institute of Applied Systems Analysis (1981).

# Biographical notes

**Theodore Beresowski** is an American engineering physicist and mechanical engineer currently heading Unesco's Energy Development and Co-ordination Section.

**Edgar J. DaSilva** is an Indian microbiologist who has worked for the United Nations Environment Programme and Unesco since 1977.

**David Oakley Hall** is interested in solar energy, photosynthesis and iron-sulphur proteins. Dr Hall is professor of biology at King's College, University of London, United Kingdom.

**J. Gururaja,** of India, heads the New Energy Sources Section, Department of Science and Technology, New Delhi, India.

The **International Fusion Research Council**'s contributors include: C. Braams (Netherlands), M. Brennan (Australia), B. Brunelli (Italy), G. von Giercke (Federal Republic of Germany), K. Husimi (Japan), E. Kintner (United States), B. Lehnert (Sweden), D. Palumbo (Commission of the European Communities), R. Pease (United Kingdom), M. Trocheris (France) and B. Kadomstev (Union of Soviet Socialist Republics). They were assisted by the following members of the International Atomic Energy Agency (IAEA); J. Phillips, A. Belozerov (scientific secretary) and H. Seligman. The report was submitted in final form to Sigvard Eklund, Director-General of the IAEA.

**Vladimir A. Kouzminov** is a Soviet specialist in mechanical and power engineering assigned to Unesco's Energy Development and Co-ordination, where he has worked since 1978.

**James McDivitt** has been director, since 1979, of Unesco's Division of Technological Research and Higher Education; he is a Canadian mining engineer with long experience in the development problems of South-East Asia.

**Alan McDonald,** an American, has been associated with the preparation of future energy scenarios at the International Institute for Applied Systems Analysis (IIASA), Laxenburg, Austria; he specializes in energy conservation and public policy.

**Tom Mikkelsen** is a Danish medical researcher concerned with problems of bacterial endotoxins, immunity and evolution; he is attached to the Rigshospitalet (State Hospital), Copenhagen.

**Arcot Ramachandran,** of India, is currently executive director of the United Nations Programme on the Habitat, with headquarters at Nairobi, Kenya.

**Walter R. Schmitt** is on the staff of the Scripps Institution of Oceanography of the University of California, at LaJolla.

The **United Nations Environment Programme** was created subsequent to the United Nations Conference on the Environment (Stockholm, Sweden, 1972); it plays co-ordinating and catalytic roles in environmental monitoring, terrestrial ecosystems, environment and development, and environmental health (among other fields).

**Zoran Zaric** is a professor at the International Centre for Heat and Mass Transfer in Belgrade, Yugoslavia.

**Robin Clarke,** the editor of this volume, earned a degree in natural science at Cambridge University, United Kingdom, in 1960. He subsequently became a science editor and writer, becoming the editor first of the British magazine *Discovery* and subsequently of *Science Journal*. During 1970–71, he worked with Unesco on the preparation of a major survey of science and society called *The Scientific Enterprise, Today and Tomorrow*. Since then he has specialized in the fields of renewable energy and appropriate technology. He was English-language and appropriate-technology editor of the UNEP journal *Mazingira* during 1977–79, and is now editor of the British publication *Natural Energy*. He is also the author of several books, as well as co-editor of *Harvesting Ocean Energy*, a volume published in 1981 by Unesco.

[I] PUB 81/XI-1/A